Poster Sessions

Held at the 2024 AIChE Annual Meeting

San Diego, California, USA
27-31 October 2024

ISBN: 979-8-3313-1689-1

Printed from e-media with permission by:

Curran Associates, Inc.
57 Morehouse Lane
Red Hook, NY 12571

CURRAN ASSOCIATES INC.
proceedings
.com

Some format issues inherent in the e-media version may also appear in this print version.

AIChE
120 Wall Street, FL 23
New York, NY 10005-4020

Phone: (800) 242-4363
Fax: (203) 775-5177

www.aiche.org

Additional copies of this publication are available from:

Curran Associates, Inc.
57 Morehouse Lane
Red Hook, NY 12571 USA
Phone: 845-758-0400
Fax: 845-758-2633
Email: curran@proceedings.com
Web: www.proceedings.com

TABLE OF CONTENTS

GENERAL POSTER SESSION

735a Understanding the behavior of Zinc slurries in Zn-MnO_2 Alkaline Batteries

Devadharshini Kathan, William Mustain

Zn-MnO_2 batteries have dominated the primary battery market for several decades. Even though these batteries are well developed and provide reliable power at low cost, the performance of Zn-MnO_2 batteries is limited by the Zinc Anode due to corrosion, passivation, and decreased Zinc utilization. The Zinc anode is comprised of 60-80 wt.% Zinc powder, 20-40 wt.% KOH gelled electrolyte, and <1 wt.% of surfactants. Despite its high mass fraction, the volume fraction of Zinc is typically low, which leads to poor particle-to-particle contact. To conduct electrons, the complex charge-conducting mechanisms have been proposed that rely on the movement of charge both within and between the particles [1], which leads to slurry ionic conductivity being several orders of magnitude lower than the electronic conductivity. This results in a mass transport-limited reaction front that migrates radially inwards from the anode-separator interface towards the center of the current collector, leading to passivation of the anode, incomplete discharge, and increased gassing. In addition to this, corrosion is another challenge encountered in the Zinc anode where highly alkaline gelled electrolyte (pH ~15) leads to significant hydrogen evolution on the Zinc particle interface.

In this study, a novel methodology is reported to quantify the contributions of the ionic and electronic conductivity in the discharge and corrosion of Zn slurry anodes. Through this new understanding, challenges in passivation and corrosion can possibly be mitigated by optimizing the composition of the slurry including Zinc weight percent, KOH concentration, and surfactant type and amount of surfactant and analyzing their behaviors using various electrochemical techniques in ex-situ and in-situ cells. The results of the presented work aim to provide new design principles for the study and utilization of slurry-based anodes for primary Zn batteries and beyond.

References

1. Faegh, T. Omasta, M. Hull, S. Ferrin, S. Shrestha, J. Lechman, D. Bolintineanu, M. Zuraw and W.E. Mustain, "Understanding the Dynamics of Primary Zn-MnO2 Alkaline Battery Gassing with Operando Visualization and Pressure Cells", J. Electrochem. Soc., 165 (2018) A2528-A2535

735c An Experimental and Modeling Study on Gas Solubility and Transport in Ionic Liquids Employed As Sweep Solvents for Membrane Contactor Reactor Applications

Mohammad Bazmi, Sheng Hu, Fei Meng, Hamid Bakhshizarinabadi, Yongchun Tang, Theodore Tsotsis, Kristian Jessen

Methanol (MeOH) is an important compound finding applications today as fuel and an important raw material in the chemical industry. Methanol is produced industrially by the catalytic conversion of syngas (a mixture of $CO/CO_2/H_2$) produced either through the steam reforming of natural gas or the gasification of coal or biomass. Methanol synthesis (MeS) is a highly exothermic, equilibrium-limited process, technically challenged by both low per-pass carbon conversions (typically, ~25%) and difficulties with heat management during reactor operation.

In place of the conventional MeS processes, our team recently proposed a novel reactive separation technology that makes use of a membrane contactor reactor (MCR)[1,2]. The process employs a non-volatile, high-temperature resistant ionic liquid (IL) as a sweep fluid to remove in situ the MeS products (H_2O and MeOH). As a result, significantly improved per-pass carbon conversions and MeOH selectivity are obtained.

For optimal reactor design and operation, it is essential to have effective models for estimating the solubility and transport properties of the MeS reaction components, including $CO/CO_2/H_2/H_2O$/MeOH (both individually and in the mixture) in the IL (in this case EMIM-BF_4) at the relevant high pressure and temperature MeS process conditions. In the present study, the solubilities of the MeS components in EMIM-BF_4 were measured experimentally using a pressure-volume-temperature (PVT) system. To model and analyze the experimental behavior, the Peng-Robinson and Soave-Redlich-Kwong equation of states (EOS) were applied to account for the non-ideality of the gas phase, while the UNIQUAC and UNIFAC activity coefficient models were used to calculate the fugacity of the liquid phase in equilibrium with the gas phase. The modeling results are compared with our experimental solubility data, and the applied models are shown to be capable of predicting the experimental behavior. In addition, the sensitivity of MeS-MCR model predictions to the accuracy of the thermodynamic models is investigated. CO_2 diffusivities were measured both in a PVT cell (constant volume diffusion experiments) and also in a high-pressure TGA (thermogravimetric analysis) system.

The measured thermodynamic and transport properties are used to optimize the performance of the membrane reactor system. These experimental data are integrated in our in-house MeS-MCR models[1] and are shown capable of predicting the experimental behavior. Currently, we are looking at other IL candidates to optimize the final sweep solvent choice based on performance, cost, chemical stability, safety, and environmental considerations.

References:

1. Zebarjad, S.F., Gong, J., Li, Z., Jessen, K., Tsotsis, T.T., “Simulation of Methanol Synthesis in a Membrane-Contactor Reactor,” In Press J. Membrane Science, 2

2. Zebarjad, F., Hu, S., Li, Z., and Tsotsis, T.T., “Experimental Investigation of the Application of Ionic Liquids to Methanol Synthesis in Membrane Reactors,” *Ind. Eng. Chem. Res.*, 58. 11911, 2019.

735d Silicon-Vanadium Carbide Based Composite Anodes and Liquid Electrolytes for High-Performance Li-Ion Batteries

Rohit Choudhury, Vinod Janardhanan, Praveen Meduri

As the global demand for energy continues to rise, it is imperative to advance renewable energy technologies and integrate them with energy storage systems to ensure consistent power supply. Over the past few decades, significant research efforts have been directed towards enhancing the energy density of lithium-ion batteries (LIBs) [1]. Currently, graphite is the standard anode material in commercial LIBs. However, its relatively low theoretical capacity (372 mAh g^{-1}) limits its application in high-energy devices. Silicon (Si), on the other hand, presents a promising alternative due to its high theoretical capacity (4200 mAh g^{-1}), abundant availability, and suitable lithium uptake potential (0.4 V vs Li/Li+) [2]. However, the practical application of silicon as an anode material is hindered by its substantial volume expansion (400-500%) during operation [3].

In the present investigation, we intend to primarily mitigate the volume expansion issue of silicon (Si) leading to the fabrication of high-performance LIBs. For realising this, Si, in micro (commercial) and nano size (synthesized using magnesiothermic reduction) have been used. At first, Si particles are amalgamated with graphite to form a finely ground mixture (Si-G). This mixture is then combined with a polymer solution to produce pyrolytic carbon, which serves to buffer the mechanical stresses induced by the volumetric expansion of Si. This process also prevents the direct contact of Si with the electrolyte, thereby stabilizing the solid electrolyte interphase (SEI). The incorporation of a dopant enhances the inherent conductivity of the material, leading to an increase in capacity by facilitating the transport of electrons and lithium ions. To further enhance the performance of the composite, synthesized Mxene vanadium carbide (V_2C) is introduced, which promotes charge transport among the Si particles and helps to increase the lithium-ion diffusion through the material. Density Functional Theory (DFT) simulations are performed to show that addition of V_2C increases the mobility of Li+ ions through the structure. The morphology and structure of the composite are examined using a range of techniques, including SEM-EDS, TEM, XRD, Raman, and XPS spectroscopy. These analyses reveal the intricate structure of the materials with a uniform distribution of Si, C, and Mxene sheets, with the sheet architecture being preserved.

We also tested different liquid electrolytes (along with additives) with a Si-C composite material as anode. Galvanostatic Charge/Discharge (GCD) tests, thermal analysis at different temperatures and other electrochemical characterizations are performed to study the electrolytes' behaviour and how their individual components affect the overall electrochemical performance of the cell.

The Si-Mxene composite, when employed as an anode, demonstrates exceptional electrochemical performance. This is substantiated by its remarkable lithium storage specific capacity of 2003 mAh g^{-1} (calculated on the basis of Si weight), for micron Si and ~3100 mAhg^{-1} (for nano Si) which it retains even after prolonged cycling at a high current density. The volume change was reduced from 420% to 150% (for micron Si) and 33% (for nano Si). Furthermore, the composite exhibits superior rate performance, delivering a specific capacity of 2439 mAh g^{-1} at a 10 C rate, underscoring its suitability for high-power applications. Additionally, the electrode showcases a low charge transfer impedance and rapid electron transport, contributing to its enhanced electrochemical performance. This improvement can be attributed to the unique structure of the composite, which comprises Mxene and doping. This structure significantly augments the diffusion of Li ions and the number of active sites. Consequently, this study presents a viable strategy for the development of high-capacity lithium-ion batteries (LIBs) based on silicon.

References

1. Liu, L. Kang, J. Hu, E. Jung, J. Zhang, S.C. Jun, Y. Yamauchi, ACS Energy Lett. 6 (2021) 3011–3019.
2. Zhou, Y. Liu, C. Du, Y. Ren, T. Mu, P. Zuo, G. Yin, Y. Ma, X. Cheng, Y. Gao, Journal of Power Sources. 381 (2018) 156–163.
3. Yi, F. Dai, M.L. Gordin, S. Chen, D. Wang, Advanced Energy Materials 3 (2013) 295-300.

735e Relationship between the Polymer Blend and Antimicrobial Activities Against *Staphylococcus Aureus*

Linh Doan, Khoa Tran

The findings from Pareto charts, main effect plots, and interaction plots demonstrate the importance of polymer concentration. Increasing concentration improves the inhibition percentage and decreases the MIC_{50}. However, the primary factor that influences these changes is chitosan (CS). Additionally, the interaction between CS and PVP, along with other polymers, plays a crucial role in achieving better antimicrobial effects. These results enhance our understanding of the antimicrobial properties of the studied polymers and offer valuable insights for developing effective antimicrobial formulations. The MIC_{50} value of M1–M16 was at a polymer percentage of 12.5%. At 12.5% polymer percentage, with the limits of [PVA], [PEG], and [PVP] being 0.002–0.004 g/mL and [CS] being 0.001–0.002 g/mL, using the 2-level full factorial method, the inhibition percentage is equal to 174.1 − 27,812 PVA − 18,561 PVP − 25,960 PEG − 38,752 CS + 9,263,047 PVA*PVP + 10,430,763 PVA*PEG + 15,397,157 PVA*CS + 7,088,313 PVP*PEG + 7,841,221 PVP*CS + 14,228,046 PEG*CS − 3,367,292,860 PVA*PVP*PEG − 5,671,998,721 PVA*PVP*CS − 6,619,041,275 PVA*PEG*CS − 3,917,095,529 PVP*PEG*CS + 2,273,661,969,470 PVA*PVP*PEG*CS. Theoretically, the most economical concentrations of PVA, PVP, PEG, and CS are 0.002, 0.002, 0.002, and 0.001 mg/mL at a concentration of 12.5% to reach an inhibition percentage of 99.162%, which coincides with the MBC value.

735f Ionic Drug Transport in Charged Nanostructured Biosponge Polymers to Reduce Chemotherapy Toxicities

Hee Jeung Oh

Cancer is becoming the leading cause of death in most developed nations. Despite efforts to develop targeted and personalized cancer therapeutics, dosing of the cancer chemotherapeutics is limited by toxic side effects. Typically, more than 90% of the injected drug is *not* trapped in the target organ and bypasses the tumor to general circulation, causing toxicities in distant locations. In the context of reducing the toxicity of chemotherapy, we have designed charged nanostructured biosponge adsorbers for capturing ionic chemotherapy drugs before they spread through the body. Doxorubicin, a widely used chemotherapy drug with significant toxic side effects, is a model drug. The charge concentration and water uptake in the biosponge adsorbers are systematically changed to achieve high binding capacity and fast kinetics. We aim to understand the mechanism of ionic drug transport in these biosponge adsorbers and establish the design principles for multifunctional biosponge polymers. Using a swine model, our initial design enables the capture of 69 % of the administered drug without any adverse effects. The proposed approach may help patients fight cancer.

735g Design and Optimization of Carbon-Coated Silica Nano Spring System for Solar-Driven Evaporation for Water Desalination

Mounika Chevula, Sharad Puri, Daniel Lippert, Dongjin Seo, David N. McIlroy, Seokjhin Kim

Solar driven evaporation offers a viable solution to global freshwater scarcity by utilizing abundant solar energy as the sole and sustainable energy source, reducing high energy consumption in conventional methods. Nevertheless, many solar evaporators experience low water evaporation rates or rely on expensive light absorbers and intricate processes. This study introduces a high-efficiency solar evaporator for treating brackish and produced water in water-scarce regions. The system integrates dense silica 1D nanosprings (NS) grown via the vapor-liquid-solid (VLS) mechanism on an alumina ceramic rod (ACR). The incorporation of 1D NS on the ACR results in enhanced evaporating surface area and induces capillary action, leading to a rise in the water level. Subsequently, a conformal layer of thin carbon, known as Graphene from the University of Idaho Thermolyzed Asphalt Reaction (GUITAR), is coated using atmospheric pressure chemical vapor deposition (APCVD). Consequently, the proposed system exhibits a high evaporation rate of 3.64 kg m^{-2} h^{-1} for 150 minutes continuous evaporation test under 1 sun irradiation. This study offers novel insights into the design and optimization of a nanostructure-based, high-efficiency solar evaporator, opening new perspectives in the field.

735h Engineering Silk Fibroin-Based Ionogels of Tuneable Physicochemical and Rheological Properties

Talia Shmool, Laura K. Martin, Andreas Jirkas, Richard P. Matthews, Anna P. Constantinou, Devkee M. Vadukul, Theoni K. Georgiou, Francesco A. Aprile, Jason P. Hallett

Ionogels are captivating materials offering high transparency, electroconductivity, and thermal and structural stability.[1,2] Conventionally, ionogels are comprised of ionic liquids (ILs), consisting of cations and anions, confined within polymer networks. Typically, ionogels require complex chemistry, show high viscosity, poor biocompatibility, and low biodegradability.[2] In this work, we develop a rational formulation design approach for facile construction of novel thermoresponsive ionogel formulations, addressing the issues of conventional ionogels for pharmaceutical applications.[1] The ionogels were comprised of biocompatible choline-based ILs, silk fibroin, and traditional pharmaceutical excipients, including tween 20, histidine, trehalose, sucrose, glycerol, and polyvinylpyrrolidone (PVP). We utilized the ILs, choline acetate ([Cho][OAc]), choline dihydrogen phosphate ([Cho][DHP]), and choline chloride ([Cho][Cl]). These ILs have previously been shown to lower formulation viscosity, increase gel strength, tune the gelation properties of diverse systems, and improve drug solubility and storage stability.[3,4] We found that formulations lacking IL failed to show gels formation; yet formulations including IL resulted in ionogel formation. Visual tests, transmission electron microscopy, and confocal reflection microscopy, demonstrated the formation of distinct ionogels, also improving the solubility of the poorly soluble antiepileptic drug, phenobarbital. Overall, [Cho][DHP] and [Cho][OAc]-ionogel formulations exhibited relatively lower hydrodynamic diameter (D_h) and more negative zeta potential values compared to the [Cho][Cl]-ionogel formulations. Furthermore, formulations lacking IL exhibited relatively higher D_h and polydispersity index values and less negative zeta potential values. This indicated that inclusion of IL reduced aggregation propensity, and likely, improved system structural stability. Under 25 °C storage, overall, the [Cho][DHP]-ionogels and [Cho][OAc]-ionogels formed relatively rapidly, shortest following 1 day. The [Cho][Cl]-ionogels required longer formation periods, extended to 32 days. Addition of PVP to each formulation reduced the Ionogel formation period to 14 days. Rheological measurements at 25 °C showed the formation of diverse ionogels of strengths between 18 and 642 Pa. Of the systems examined, [Cho][Cl]-ionogels exhibited the lowest strengths. Furthermore, addition of PVP increased the strengths of the ionogels. This was most prominent for [Cho][OAc]-ionogels and [Cho][DHP]-ionogels also including glycerol. These exhibited maximum storage modulus (G') values of 167 and 642 Pa, respectively. Additionally, we found that formulations showing lower G' values also required longer ionogel formation time, and disaccharide identity was interchangeable. Finally, heating the fresh aqueous formulations including IL, from 25 to 60 °C, resulted in the formation of thermoresponsive ionogels between 35 and 41 °C. Addition of PVP resulted in the formation of ionogels between 44 and 49 °C. Conversely, formulations lacking IL presented as cloudy solutions upon heating. Based on the experimental findings and density functional theory calculations, we propose that the dihydrogen phosphate anions act as hydrogen bond donors and acceptors, facilitating a network of strong intermolecular hydrogen bonding interactions. This explains the rapid [Cho][DHP]-ionogel formation period. Accordingly, since the acetate anions act solely as hydrogen bond acceptors, relatively weak ionogels formed. Moreover, inclusion of the chloride anions resulted in the weakest intermolecular hydrogen bonding interactions explaining the delayed formation of [Cho][Cl]-ionogels. We suggest that the addition of glycerol and PVP to the formulations resulted in strong intermolecular hydrogen bonding and macromolecular crowding, contributing to ionogels of greater strength, structural and energetic stability. As increasingly advanced ionogel designs are explored, applying our rational development strategy enables the production of ionogels which can be formulated for precise control of multiple unique ionogel features. We believe, these could potentially fill niche pharmaceutical applications.

References

1. Shmool, T.A.; Martin, L.K.; Jirkas, A.; Constantinou, A.P.; Vadukul, D.M.; Aprile, F.A.; Georgiou, T.K.; Hallett, J.P. Unveiling the Rational Development of Stimuli-Responsive Silk Fibroin-Based Ionogel Formulations. *Chem. Mater.* **2023**, *35*, 5798–5808.

2. Kopilovic, B.; e Silva, F.A.; Pedro, A.Q.; Coutinho, J.A.P.; Freire, M.G. Ionogels for Biomedical Applications. In *Nanotechnology for Biomedical Applications*; Gopi, S., Balakrishnan, P., Mubarak, N.M., Eds.; Materials Horizons: From Nature to Nanomaterials; Springer: Singapore, 2022; pp 391–425.

3. Shmool, T.A.; Constantinou, A.P.; Jirkas, A.; Zhao, C.; Georgiou, T.K.; Hallett, J.P. Next Generation Strategy for Tuning the Thermoresponsive Properties of Micellar and Hydrogel Drug Delivery Vehicles Using Ionic Liquids. *Polym. Chem.* **2022**, *13*, 2340–2350.

4. Nahar, Y.; Horne, J.; Truong, V.; Bissember, A.C.; Thickett, S.C. Preparation of Thermoresponsive Hydrogels Via Polymerizable Deep Eutectic Monomer Solvents. *Polym. Chem.* **2021**, *12*, 254–264.

735i Squid-Inspired Materials with Tunable Heat-Managing Properties

Aleksandra Strzelecka, Panyiming Liu, Sanghoon Lee, Alon Gorodetsky

Advanced heat-managing technologies have attracted widespread attention due to their potential to enhance personal comfort and reduce building energy consumption. Within this context, the development of systems that can merge the advantageous characteristics of both active and passive operating modes has proven particularly attractive. Recently, we have developed squid skin-inspired composite materials with user-tunable heat-managing properties, which feature several key advantages. First, the composites feature the ability to modulate infrared transmittance by > 20-fold and heat fluxes by > 30 W/m^2 upon actuation with strain (1). Second, the composites' adaptive infrared and dynamic thermoregulatory properties can be computationally predicted via straightforward models (1,2). Third, the composites can be modified to endow them with breathability, washability, and fabric compatibility (3). Fourth, the composites can be scalably manufactured over areas as large as 0.3 m^2 by using standard techniques (1). Our combined findings may enable applications of the thermoregulatory materials in multifunctional wearables, food packaging, infrared camouflage, soft robotics, and biomedical sensing.

References:

1. Badshah, M. A., Leung, E. M., Liu, P., Strzelecka, A. A., Gorodetsky, A. A. Scalable manufacturing of sustainable bioinspired packaging materials with tunable heat-managing properties. *Sustainability,* **5**, 434–443 (2022).
2. Liu, P., Leung, E. M., Badshah, M. A., Moore, C., Gorodetsky, A. A. Structure-Function Relationships for Squid Skin-Inspired Wearable Thermoregulatory Materials. *APL Bioeng. 7, 046111* (2023).
3. Lee, S., Leung, E. M., Badshah, M. A., Strzelecka, A. A., Gorodetsky, A. A. Manufacturing of Breathable, Washable, and Fabric-Integrated Squid Skin-Inspired Wearable Thermoregulatory Materials. (In Revision, 2024)

735j Biological Approaches for Beverage Wastewater Treatment and the Production of Single Cell Protein: A Life Cycle Assessment

Yuming Wen, Qiang Guo, Lifeng Li, Zulfida Mohamad Hafis Bin Mohd Shafie, Yan Zhou, Raymond Lau, Chi-Hwa Wang

Wastewater from beverage industry has become a global challenge. A bio-platform was proposed recently to utilize beverage wastewater to culture microalgae and bacteria. As such, wastewater can be treated and highly valued fish feeds can be produced simultaneously. The proposed platform consists of three different scenarios according to the different species that are cultured, including microalgae, phototrophic bacteria, and heterotrophic bacteria. Life Cycle Assessment was undertaken in this study to evaluate the environmental impacts of these three scenarios. For each scenario, stages such as cultivation, harvesting, wastewater treatment, separation, drying, and byproduct management were considered. The results show that when the concern is to increase the production of fish feeds, the scenario with microalgae should be maximized, while if the goal is to increase the amount of the wastewater that needs to be treated, the process with heterotrophic bacteria should be maximized. Suggestions were also given to replace parts of energy for cultivation and separation with solar PV and solar tower to mitigate environmental impacts.

735k Circularizing Animal Wastewaters: Spatial Mapping of Swine Wastewater and Associated Resource Recovery Potential in the United States

Madison Kratzer, Ahteshamul Haq, Prathap Parameswaran, Vikas Khanna

The circular economy framework is aimed at meeting the resource demands of the anthroposphere while minimizing the necessity for demand reduction strategies within our economy. A fundamental principle of the circular economy involves prolonging the residence time of resources in the economy through approaches like recycling and resource recovery. Animal wastewater represents a resource-rich stream containing significant amounts of biological oxygen demand (BOD), nitrogen, and phosphorus that can be reclaimed. Specifically, the trend towards large, concentrated animal feeding operations (CAFOs) within the swine industry makes its wastewater an appealing target for resource recovery due to its strength and relative consistency. Swine wastewater can be utilized for the production of biogas, fertilizers, and value-added chemicals through advanced anaerobic biotechnologies. However, systems-level assessment is needed to quantify the spatial availability of swine wastewater as well as its resource recovery.

This work will present creation of an account of swine wastewater in the United States based on publicly available data. This includes data regarding head count, farm type, and farm location at both state and county levels. Furthermore, this study presents a spatial analysis of the transfer of swine wastewater created by the movement of swine to different farms through their life stages. This analysis is part of an effort to conceptualize and consider swine wastewater as an economic commodity. The movement of swine across life cycle stages is examined using network analysis and optimization techniques, with a focus on minimizing the distance traveled by swine to match supply and demand of swine farms for different life stages. Finally, the study presents an assessment of the total value of recovered energy and nutrients and quantification of greenhouse gas emission reduction resulting from the recovered resources from swine wastewater. Preliminary findings show that swine wastewater is concentrated in the corn belt allowing for potential of combining multiple wastewater streams for more efficient treatment using an AnMBR technology. Swine wastewater in the United States was found to have over 6,000 MW of recoverable energy, enough to theoretically power New York City if all were collected.

735l Advancing Plant Cytochrome P450 Functionality in Yeast

Shanhui Xu, Yanran Li

Plant natural products (PNPs) are important for drug discovery, yet their complex structures and low abundance pose challenges for sourcing. Cytochrome P450s (CYPs), which catalyze various oxidation in plants, are key players in PNP biosynthesis. While yeast fermentation has emerged as a promising alternative for PNP production, achieving efficient functional reconstitution of CYPs in yeast remains a major challenge. Traditional strategies for enhancing CYPs, such as protein engineering and electron transfer optimization, do not consistently yield effective results and are context dependent. Recently, the role of the microenvironment, especially membrane composition, was recognized in supporting CYP function. In our study, we identified plant membrane steroid binding proteins as a general enhancer for CYP activities in baker's yeast. By investigating the function of these steroid binding proteins in yeast, we illuminates the scaffold protein-mediated regulation of CYP activities. Through targeted engineering of relevant microenvironment, we successfully enhanced CYP activities in yeast. This work provides new insights for improving CYP activities in yeast.

735n Hydrogen from Air ($SAWH_2$): A Decentralized Sorption Based Atmospheric Water-Hydrogen Production Device

Joseph Phelim Mooney

Hydrogen from Air ($SAWH_2$): A Decentralized Sorption Based Atmospheric Water-Hydrogen Production Device

Joseph P. Mooney[1,2*], *Chad T Wilson*[1], *Omer Refet Caylan*[1], *David Keisar*[1], *Carlos Díaz Marin*[1], *Bachir El Fil*[1*]

[1]Department of Mechanical Engineering, Massachusetts Institute of Technology, Cambridge, MA 02139, USA

[2]School of Engineering, University of Limerick, V94 T9PX, Ireland

jpmooney@mit.edu ; belfil@mit.edu

The escalating challenges posed by climate change and water scarcity necessitate innovative solutions in the energy sector. A key component of this innovation is the adoption of hydrogen as a clean, versatile alternative to traditional energy sources. Hydrogen, particularly when produced through the process of electrolysis using renewable energy sources like solar power, stands at the forefront of the transition towards a more sustainable and low-carbon future. This process, known as green hydrogen production, is gaining traction as a viable method for clean energy generation. However, the path to fully harnessing green hydrogen's potential is fraught with obstacles, including the significant demand for water and the limitations in production efficiency due to the reliance on natural sunlight alone. The production of green hydrogen, especially in arid or water-scarce regions, faces the critical challenge of water availability. This issue underscores the need for green hydrogen production methods that are not only efficient but also sustainable in their water use. Furthermore, green hydrogen systems often struggle to surpass energy conversion efficiencies of 10%, hindered by suboptimal energy capture, oxygen and hydrogen bubble transport, conversion processes, and losses during electrolysis. Overcoming these efficiency and sustainability barriers is crucial for the scalability and broader application of green hydrogen as a reliable energy carrier. Sorption-based atmospheric water harvesting (SAWH) emerges as an energy-efficient solution for obtaining ultra-pure water off-grid, crucial for applications such as green hydrogen production. This technology harnesses ambient thermal energy for water adsorption and desorption, minimizing the need for external power sources. Additionally, SAWH can be integrated into the thermal management of photovoltaic (PV) cells, enhancing their efficiency and longevity by mitigating heat stress. Furthermore, by expanding the usable wavelength range of the solar spectrum, SAWH contributes to a marked increase in the efficiency of hydrogen production, leveraging the full solar spectrum. In this work we propose a decentralized green hydrogen production ($SAWH_2$) device that integrates water adsorption-desorption processes with solar energy conversion, offering a dual solution to the pressing needs for energy and clean water. This approach not only enhances the efficiency and sustainability of hydrogen production but also harnesses atmospheric humidity as a water source. By improving water resource utilization and energy conversion efficiency, the system paves the way for a scalable and sustainable method of solar-powered green hydrogen production.

The system comprises a sorbent-based atmospheric water harvesting material (desiccant) strategically deployed either as a coating on a photovoltaic (PV) cell or within a separate dedicated unit, a hydrogen production cell, and a hydrogen storage unit. During water adsorption, the desiccant absorbs water vapor from the surrounding air, either passively or actively. Solar energy is used to produce electricity from the PV cell while simultaneously desorbing water from the desiccant, which is then employed for water splitting via electrolysis of water. Using our high-performance low-cost polyacrylamide Lithium chloride (PAM-LiCl) hydrogel sorbent (1.79 and 3.86 gg^{-1}at relative humidity (RH) of 30% and 70%, salt loading of 15-20 gram of salt per gram of polymer, and at a cost of <0.1 $/kg of material, respectively). Preliminary results show a device water production capacity of ~2L/m^2/day and a hydrogen production capacity of 5 mL/min. We will also present a heat and mass transfer analysis to optimize the system level performance of this device, such as PV to Proton Exchange Membrane (PEM) scaling, sorbent to hydrogen production comparisons as well as novel insights into passively replenishing the PEMs surface while maintain adequate pressure between the anode and cathode. We increase hydrogen production enhance efficiency while lowering the overall cost. Due to findings that the device is PV-PEM limited, we also demonstrate that our device can supply both energy and potable fresh water. Our device presents an alternative solution to green hydrogen production when presenting a starting point to getting closer to the DOE's target levelized cost of hydrogen (LCOH) of $1/kg to make it more competitive and economically viable in the broader energy landscape. Our $SAWH_2$ device is also expected to have an efficiency enhanced by >40% as compared to state-of-the-art devices. The implications of this technology extend far beyond its immediate environmental benefits. By providing both energy and water, two fundamental components for human health and prosperity, the system has the potential to transform communities worldwide. Its decentralized and affordable nature makes it particularly suitable for remote and arid regions, where access to clean water and reliable energy sources is often limited. Applications range from supplying clean drinking water and powering agricultural projects to enabling local energy production and supporting healthcare facilities. Additionally, this technology could play a crucial role in disaster relief operations, where immediate access to clean water and energy is critical.

735o Exploring the Origin of Hysteresis in Plastic Crystal Compounds for Engineering Pressure Tunable Thermal Energy Storage

Chase Somodi, Lindsey Ford, Kristin McCormick, Daniel Tabor, Emily Pentzer, Patrick Shamberger

Plastic crystals are a class of small, globular molecules that exhibit promising characteristics for use as tunable thermal energy storage (TES) materials. At low temperature or high pressure, plastic crystal molecules adopt regular translational and orientational ordering, while at higher temperatures or lower pressures, the bonds between adjacent molecules begin to break and molecules begin to rotate about their centers of mass. This transition from rotational order to disorder has a large associated isothermal entropy and adiabatic temperature change. The magnitude of this isothermal entropy and adiabatic temperature change makes plastic crystals a promising class of alternatives to environmentally harmful hydrofluorocarbons (HFCs) used in modern day refrigeration and heat pump cycles. This also gives them great potential for use as pressure-tunable TES. However, the relatively large degree of undercooling, on the order of 10-20 °C, even the best plastic crystals exhibit greatly impacts the overall cyclic efficiency. Understanding the rate-dependent phase transition mechanisms of the plastic crystal transition could help to formulate a method for reducing the undercooling. Ultimately, this could propel plastic crystals forward as next generation solid-state refrigerants or TES materials, helping to reduce the effects of man-made climate change and the energy grid demand.

In the present work, we utilize optical microscopy and calorimetric observations and analyses to investigate the kinetics in a model plastic crystal, neopentyl glycol. On heating, we demonstrate a process in which domains of the high symmetry crystal structure preferentially nucleate along grain boundaries. Subsequent growth of the isotropic high symmetry phase limits this heating transformation, which differs from the case on cooling. Observations through both methods suggest a nucleation limited cooling transformation. The metastable plastic crystal state persists well below the expected transition temperature until one domain of the ordered crystal nucleates, upon which a rapid transformation occurs. We reveal the impact of cycling on the microstructure and find that despite its associated increase in grain boundary density, the asymmetric transformation kinetics remain conserved. Lastly, we explore the magnitude of the thermal hysteresis through both experimental techniques, and discuss methods for reducing this hysteresis in future studies.

Building on this understanding, we explore the relationship between lattice correspondence between the ordered and plastic crystal structures and the thermal hysteresis in a variety of plastic crystal systems. We aim to demonstrate that the magnitude of transformation strain associated with the structural transition between the ordered and plastic crystal phases plays a critical role in the magnitude of thermal hysteresis. Through temperature dependent X-ray diffraction, we evaluate the crystal structures of the ordered and plastic crystal phases and utilize the crystal structures to determine lattice disregistry and transformation strain, serving as proxies for the strain energy at the interface between parent and daughter phases. Further, we use calorimetry to quantify the magnitude of undercooling associated with the plastic crystal transition in a variety of compounds. Combining the lattice disregistry and undercooling calculations, we demonstrate that phases which have some degree of lattice correspondence (i.e. small lattice disregistry) between the ordered and plastic crystal phase exhibit smaller degrees of undercooling. In contrast, molecules that share little to no similarity between the crystal structures of their ordered and plastic crystal phases have large degrees of undercooling, contributing to the large thermal hysteresis observed through calorimetry. Lastly, we investigate compounds with relatively low to moderate degrees of undercooling using polarized light microscopy to identify differences in kinetics for the forward and reverse transformations. We discuss the relationship these optical observations have to the crystal structure and lattice correspondence, as well as their implications on the thermal hysteresis of the plastic crystal compounds.

Utilizing the information revealed through the initial kinetics study on the model plastic crystal neopentyl glycol, as well as the study on lattice correspondence in a variety of plastic crystals, we conclude by discussing pathways for materials development. More specifically, we discuss methods for engineering low-hysteresis plastic crystal compounds for use in barocaloric refrigeration systems, or for use as thermal energy storage mediums.

735p Reactive Wetting between Gallium-Based Liquid Metals and Cu, Ni, Ti, and Au Substrates

Lucas Oelkers, Patrick Shamberger

Growing global energy demand, coupled with the urgent need to achieve net-zero carbon emissions by 2050, requires revolutionary advances in energy storage and management systems. Researchers have long utilized conventional phase change materials (PCMs), such as paraffins and salt hydrates, to increase thermal energy storage and absorb peak energy loads. However, these materials lack sufficient thermal conductivity to achieve sufficient thermal charge and discharge rates (greater than 1 W/cm^2), especially for use in concentrated solar receivers, and high-power transient electromagnetic systems [1], [2], [3]. As a strategy to quantify the relative capacity of a PCM to rapidly charge or discharge thermal energy, a figure of merit (FOM) has been derived from the analytical solution to the 1D two-phase Neumann—Stefan problem which relates the rate of thermal energy storage to specific PCM thermophysical properties [4]. This performance metric suggests that low melting point liquid metals exhibit 1 to 2 orders of magnitude greater cooling capacity compared to conventional PCMs when utilized in heat sinks with similar boundary conditions [5]. Among these, gallium-based liquid metals (GBLMs) represent one candidate metallic PCM system of great interest due to their following material properties: 1) large thermal conductivities, greater than 30 W/m/K [4]; 2) large volumetric latent heat of melting, greater than 400 MJ/m^3 [6]; and 3) low melting point temperatures, less than 30 °C [6].

Despite this interest, the interfacial characteristics of GBLMs with thermally conductive metallic heat exchanger surfaces or containment vessels are underdeveloped, and have previously demonstrated problematic characteristics including (1) formation of brittle intermetallic phases which can spall off, and dissolution of gallium and subsequent embrittlement of metallic substrates, such as Al alloys [7]; (2) depletion and composition modulation of GBLMs due to selective diffusion of GBLMs elements into the metallic substrate and the formation of intermetallic phases; and (3) the large surface tension of GBLMs, which generally exceeds 700 mN/m [8], thus inhibiting wetting in metallic surfaces; and (4) the near-instantaneous formation of oxide layers on GBLMs when handled in the air that pins the liquid's fluidity [9].

This study is motivated by the hypothesis that there exists a quantifiable maximum intermetallic layer thickness due to a self-limiting intermetallic reaction that also promotes reactive wetting (a progressive decrease in contact angle and surface tension over time due to the formation of highly wettable intermetallic phases between the GBLM and the metallic substrate), see Fig. 1. Cu, Ni, Au, and Ti, metallic substrates were chosen to represent a range of reactivity and thermal conductivities. The GBLM's contact angle and baseline length (the baseline length being the droplet's diameter that contacts the interface) was measured via a dynamic contact angle meter [10]. Additionally, GBLM droplets were placed on the substrates for various times and temperatures. The passivating nature and stability of the intermetallic reactions were discovered using back-scatter electron microscopy and measurement of diffusion profile thicknesses in polished cross-sections by wavelength dispersive spectroscopy. The activation energy and time exponents of these reactions will be calculated. We used X-ray diffraction and X-ray photoelectron spectroscopy to analyze the chemical composition of the intermetallic layer. Perturbations of the system included using various bulk metallic substrate compositions, thicknesses of thin-film metallic substrates, and surface roughness of the sputter-deposited thin-film substrates. Through a systematic exploration of the interfacial dynamics governing GBLM-metallic substrate interactions, this project has the potential to increase the technology readiness level for utilizing GBLMs in metallic thermal management systems.

References:

[1] Woods, J., and et al., 2021, "Rate capability and Ragone plots for phase change thermal energy storage," Nature Energy, 6(1), pp. 295-302.

[2] Ge, H., and et al., 2013, "Low melting point liquid metal as a new class of phase change material: An emerging frontier in energy area," Renewable and Sustainable Energy Reviews, 21(1), pp. 331-346.

[3] Tripathi, B. M., and et al., 2023, "A comprehensive review on solar to thermal energy conversion and storage using phase change materials," J. of Energy Storage, 72 pp. 108280.

[4] Shamberger, P. J., 2016, "Cooling Capacity Figure of Merit for Phase Change Materials," J. Heat Transfer, 138(2), pp. 024502-024509.

[5] Hoe, A., and et al., 2020, "Conductive heat transfer in lamellar phase change material composites," Applied Thermal Engineering 178 pp. 115553.

[6] Hale, D. V., and et al., 1971, "Phase Change Materials Handbook," Lockheed Missiles and Space Co., Huntsville, AL, Report No. NASA CR-61363.

[7] Senel, E., and et al., 2014, "Liquid metal embrittlement of aluminum by segregation of trace element gallium," Corrosion Science, 85(1), pp. 167-173.

[8] Song, M., and et al., 2021, "Interfacial Tension Modulation of Liquid Metal via Electrochemical Oxidation," Adv. Intelligent Systems, 3(8), Article 2100024.

[9] Liu, T., and et al., 2012, "Characterization of Nontoxic Liquid-Metal Alloy Galinstan for Applications in Microdevices," IEEE, 21(2), pp. 443-450.

[10] Seo, K., and et al., 2015, "Re-derivation of Young's Equation, Wenzel Equation, and Cassie-Baxter Equation Based on Energy Minimization," IntechOpen.

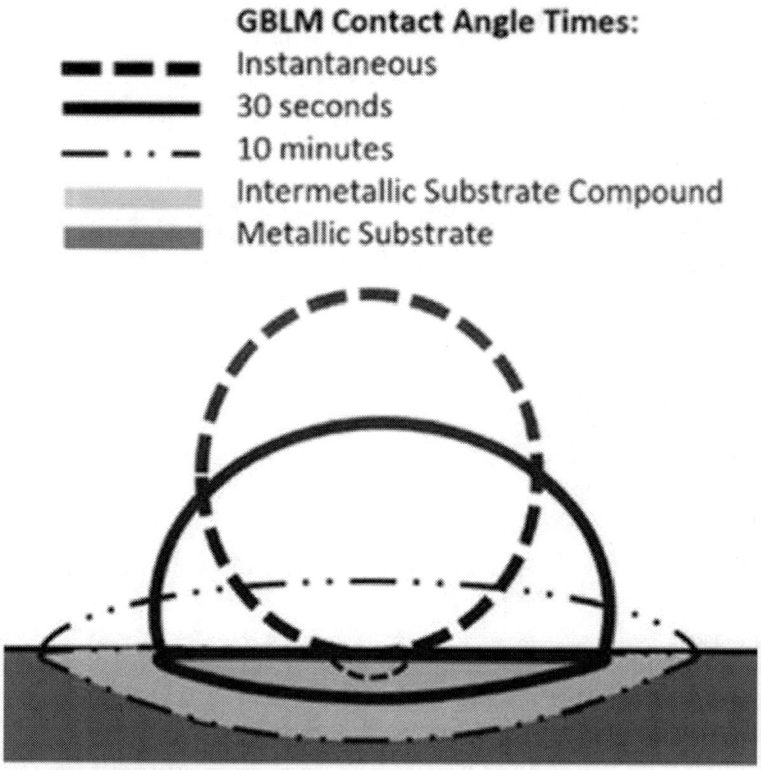

Figure 1: Theorized intermetallic layer thickness and contact angle at various times.

Development of Next-Generation Nutritionally Fortified Fermented Milk

Mahmoud Ibrahim, Negin Ammari, Gavin Walker

Abstract: Elevated blood pressure is considered to be one of the most important risk factors for cardiovascular diseases. According to the World Health Organization the leading risk factor for global mortality is hypertension. However, synthetic antihypertensive drugs are known to have side effects. Therefore, natural antihypertensive components are in high demand. Consumption of functional foods has been rising in popularity and has become a way of life for many.

The gamma-aminobutyric acid (GABA) developed in the project will be a natural alternative to synthetic molecules (ACE Inhibitors) normally used for the treatment of hypertension, heart failure and strokes. This work develops novel fermented milk that is capable to efficiently regulate blood pressure. Fermented milk rich in GABA shows significant positive impact in decreasing blood pressure. Additionally, food protein hydrolysate with antihypertensive peptide is another tool to control blood pressure. Three types of protein were selected on the basis of their antihypertensive activity including rice protein, pea protein, and casein. Also, four GABA-producing strains namely Streptococcus thermophilus -ST92, ST382, ST988, and Lactobacillus brevis-LBR01 were selected (initial data shown in image).

Strain	Antioxidant %	ACE-I %	GABA concentration (μg/mL)
Streptococcus thermophilus -ST1	29.28	20.51	17.65
Streptococcus thermophilus ST2	24.09	17.23	0.51
Streptococcus thermophilus ST3	35.34	26.01	28.5
Lactobacillus brevis-LBR01	38.03	23.31	21.67

735q Production of Curcumin Nanoparticles Via Self-Assembly in Supersaturated Solution

Parimaladevi Palanisamy, Negin Ammari, Gavin Walker

Abstract: The influence of curcumin (API) and stabilizer concentration on the size of nanoparticles was investigated using an antisolvent ratio from 1 to 80 (isopropanol-water). Initially, nanoparticles of curcumin were prepared by mixing the supersaturated solution of curcumin to the aqueous solution of monosachharides (glucose/fructose). Curcumin nanoparticles of size ranging from 120 nm to 1 μm were observed in the solution. Soluplus was later selected as a stabilizer and similar experimental procedure with various curcumin and soluplus concentrations were undertaken. Curcumin particle size was observed in a solution ratio of 1:50 ratio.

TEM analysis revealed that the curcumin nanoparticles were well dispersed in the solution (see Figure). These nanoparticles were isolated using vacuum filtration to form to micron sized particles through the adsorption of curcumin particles onto the surface of MMT clay particles. Then the adsorbed curcumin was filtered using PTFE membrane filter with 0.2 micron pore size. The concentration of curcumin in the filtrate was determined using UV-Vis spectroscopy and the drug loading and encapsulation efficiency of each sample was determined. The maximum drug loading achieved was 49 % and the maximum drug release achieved was 55 %.

To further improve drug loading and drug release spray drying was employed. In this process, drug loading of approx. 70 % and drug release of 70 % was achieved. The spray drying initially was used as a semi-continuous process. Experiments to produce curcumin nanoparticles in fully continuous mode using spray drying produces slightly larger (90 nm) curcumin nanoparticles compared to the nanoparticles (60-80 nm) in the semi-batch process.

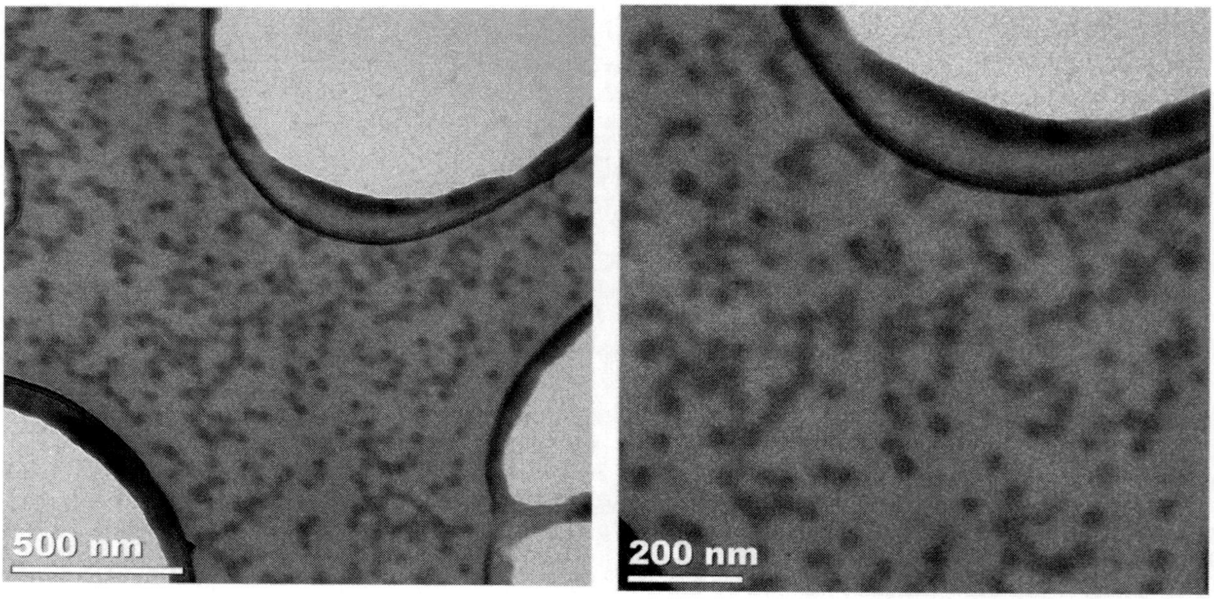

735r Characterization of a Mixed Flow Impeller in a Culture Medium

David Posadas Navarro Sr., Marco A. Ramírez Argáez V, Gabriel Ascanio

Mixing is a key unit operation that aims at the uniformity of a mixture of two or more phases. This process has applications in many industries such as chemical, petrochemical, and pharmaceutical [1].

Gas-liquid systems have been the subject of numerous studies in recent years [2, 3]. These systems have been characterized by high mass transfer, good liquid and gas mixing capacities, and longer residence times of gas in the liquid phase. These types of systems are commonly used in biological applications such as the culture of microorganisms [4,5]. In this sense, stirred tanks are used in such applications to obtain a complete homogenization of both liquid and gas phases.

To produce the required degree of uniformity inside the bioreactor the sparging and the agitation should be the best possible. The agitation inside of the bioreactor is carried out using impellers, which have been reported for different applications. These studies demonstrate that the traditional Rushton turbine impeller have some advantages over other impellers for gas-liquid mixing operations, although they exhibit drawbacks such as the formation of dead regions, poor pumping capacity, high strain rates, and high drag rate when using high viscosity fluids [6].

To overcome the deficiencies exhibited by the Rushton turbine, a hybrid impeller known as the axial radial impeller was designed and reported elsewhere [7], which is a combination of tilted blades and straight blades that creates a mixed flow. Posadas [8] studied the performance of the axial radial impeller in a gas-liquid stirred tank system using a model fluid. The results show that due to a combination of axial flow with radial flow achievement eliminates quasi-static flow zones, which resulted in better pumping, and better agitation intensity through the tank. These results show that axial-radial impellers may perform better than the traditional Rushton turbines in the fermentation processes.

In this work, the performance of a hybrid impeller called the axial-radial impeller (ARI) in a stirred tank with a culture of *Azotobacter vinelandii* was evaluated. Results are reported in terms of biomass growth, alginate production, and sucrose uptake. For comparison purposes, the biological performance of the axial-radial impeller was compared with the traditional Rushton turbine.

The bacterial strain *A. vinelandii* was grown in a modified Burk's medium which is shown in Figure 1a.

To avoid precipitation during autoclaving the solutions of and
were separated from the other medium components during sterilization. On the other hand, all components were sterilized at 121 °C for 35 minutes and were mixed in sterilized conditions.

The bioreactor consisted of a glass vessel of diameter T of 210 mm, and liquid height H of 202 mm, which provided an operating volume of 0.014 m^3. The vessel is a cylinder with a plane bottom. Four baffles having a width J of 16 mm equally spaced to minimize vortexing in the free surface were collocated. Also, a ring sparger to supply air to the vessel at a distance of 25 mm measured from the bottom of the tank. Sets of three impellers with a diameter D_a of 70 mm coaxially placed along the shaft were used, namely: Rushton turbines (RTs), and axial radial impellers (ARIs). The first impeller on the shaft was collocated at a distance of 31 mm (C_1) measured from the sparger the second impeller at a distance of 70 mm (C_2) measured from the first impeller and the third impeller at a distance of 70 mm (C_3) measured from the second impeller, and finally distance between the third impeller and the free surface was of 31 mm (C_4).

Inside the vessel both the inoculum and culture medium were poured, which were operated under the conditions described in Figure 1b. The analytical determinations such as biomass growth, alginate production, and sucrose uptake were determined using the methodology described elsewhere [9].

The effect of each one of the two impellers on bacterial growth and alginate production is shown in Fig. 3a and b. As can be seen, the major biomass growth and alginate production were obtained with the axial radial impellers. The better performance of these impellers is due to that these impellers exhibit a major and better distribution of turbulent intensity maps as reported elsewhere [8], that is these impellers transfer more efficiently the energy from the engine to fluid, which is a better environment for the growth of microorganisms. On the other hand, as can be seen in Fig. 3c the sucrose consumption is very similar between both set impellers.

Finally, the results of the kinetics growth extracted from the graphs above show that the axial-radial impellers exhibited a superior performance with a cell yield coefficient, and product yield coefficient of , and , respectively. In contrast, the values obtained by the Rushton turbines were 0.123 and 0.173, respectively. Besides, the specific growth rate was calculated. The highest specific growth rate was with a specific alginate production rate of 7.0 g/alg/g biom/h, which was obtained with the axial-radial impellers.

In this study, the biological performance of a hybrid impeller was analyzed and compared to a Rushton turbine. The most relevant conclusions are:

The axial-radial impellers show better performance compared with the Rushton turbines. The axial-radial impellers exhibit the highest parameters compared with the one obtained with the Rushton turbines, which are listed following:

1. The specific growth rate was 195.3 % higher.
2. Its cell yield coefficient was 202 % higher.
3. Its product yield coefficient was 195.3 % higher.

The better performance of the axial-radial impellers is due to that these sets of impellers demand more power than the Rushton turbines, which is reflected in a better distribution of energy through the tank.

Figure 1. Kinetics of growth of A. vinelandii with ARIs and RTs: a) Components of modified Burk's medium; b) Bioreactor operating conditions; c) Biomass growth; d) Alginate production; e) Sucrose uptake

References:

[1] Paul, E. L., Atiemo-Obeng, V. A., & Kresta, S. M. (2004). *Handbook of industrial mixing* (pp. 1-143). NY: Wiley-Blackwell.

[2] Liu, Y., Guo, J., Li, W., Li, W., & Zhang, J. (2021). Investigation of gas-liquid mass transfer and power consumption characteristics in jet-flow high shear mixers. *Chemical Engineering Journal, 411*, 128580.

[3] Forte, G., Alberini, F., Simmons, M., & Stitt, H. E. (2021). Use of acoustic emission in combination with machine learning: monitoring of gas–liquid mixing in stirred tanks. *Journal of Intelligent Manufacturing, 32*(2), 633-647.

[4] Díaz-Barrera, A., Gutierrez, J., Martínez, F., & Altamirano, C. (2014). Production of alginate by *Azotobacter vinelandii* grown at two bioreactor scales under oxygen-limited conditions. *Bioprocess and biosystems engineering, 37*, 1133-1140.

[5] Medina, A., Castillo, T., Flores, C., Núñez, C., Galindo, E., & Peña, C. (2023). Production of alginates with high viscosifying power and molecular weight by using the AT9 strain of *Azotobacter vinelandii* in batch cultures under different oxygen transfer conditions. *Journal of Chemical Technology & Biotechnology, 98*(2), 537-543.

[6] Nienow, A. W. (1996). Gas-Liquid Mixing Studies-A Comparison of Rushton Turbines with Some Modern Impellers. *Chemical Engineering Research & Design, 74*(4), 417-423.

[7] Ascanio, G., Foucault, S., & Tanguy, P. A. (2003). Performance of a new mixed down pumping impeller. *Chemical Engineering & Technology: Industrial Chemistry-Plant Equipment-Process Engineering-Biotechnology, 26*(8), 908-911.

[8] Posadas-Navarro, D., Palacios, C., Blancas-Cabrera, A., Trujillo-Roldán, M. A., Salinas-Vázquez, M., & Ascanio, G. (2022). Flow Patterns of Multiple Axial-Radial Impellers for Potential Use in Aerated Stirred Tanks. *Chemical Engineering & Technology*, *45*(5), 860-867.

[9] Peña, C., Campos, N., & Galindo, E. (1997). Changes in alginate molecular mass distributions, broth viscosity and morphology of *Azotobacter vinelandii* cultured in shake flasks. *Applied Microbiology and Biotechnology*, *48*, 510-515.

a)

Reagent	Concentration [g/L]
Sucrose	20.0
Yeast extract	3.0
K_2HPO_4	0.66
KH_2PO_4	0.16
MOPS	1.42
$CaSO_4$	0.05
NaCl	0.2
$MgSO_4 \cdot 7H_2O$	0.2
$Na_2MoO_4 \cdot 2H_2O$	0.0029
$FeSO_4 \cdot 7H_2O$	0.027

b)

Parameter	Value
Agitation rate	300 rpm
Airflow	1 vvm
pH	7.2
Temperature	29 °C

c)

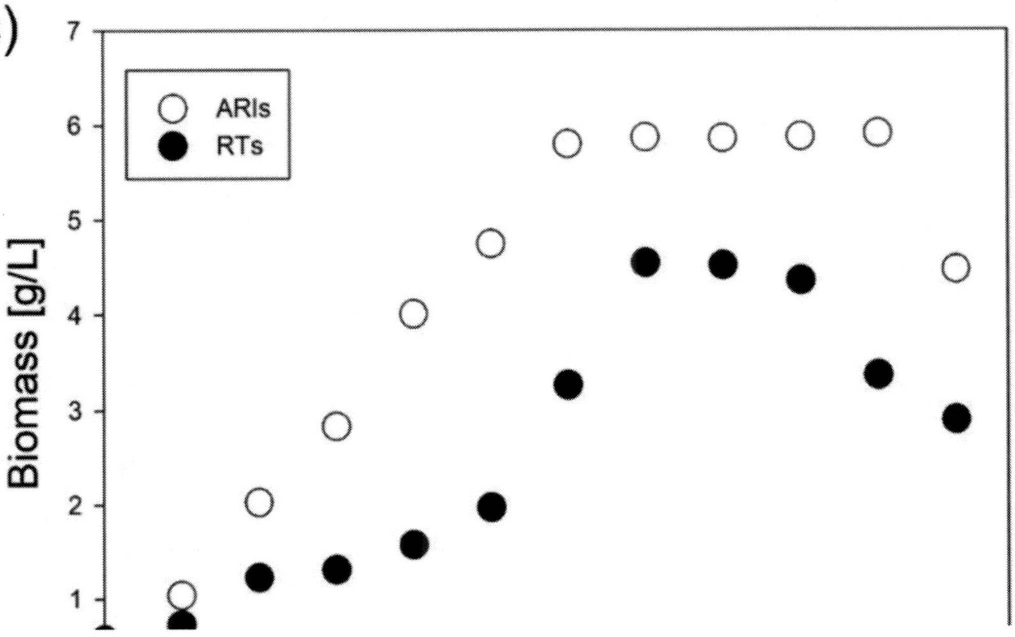

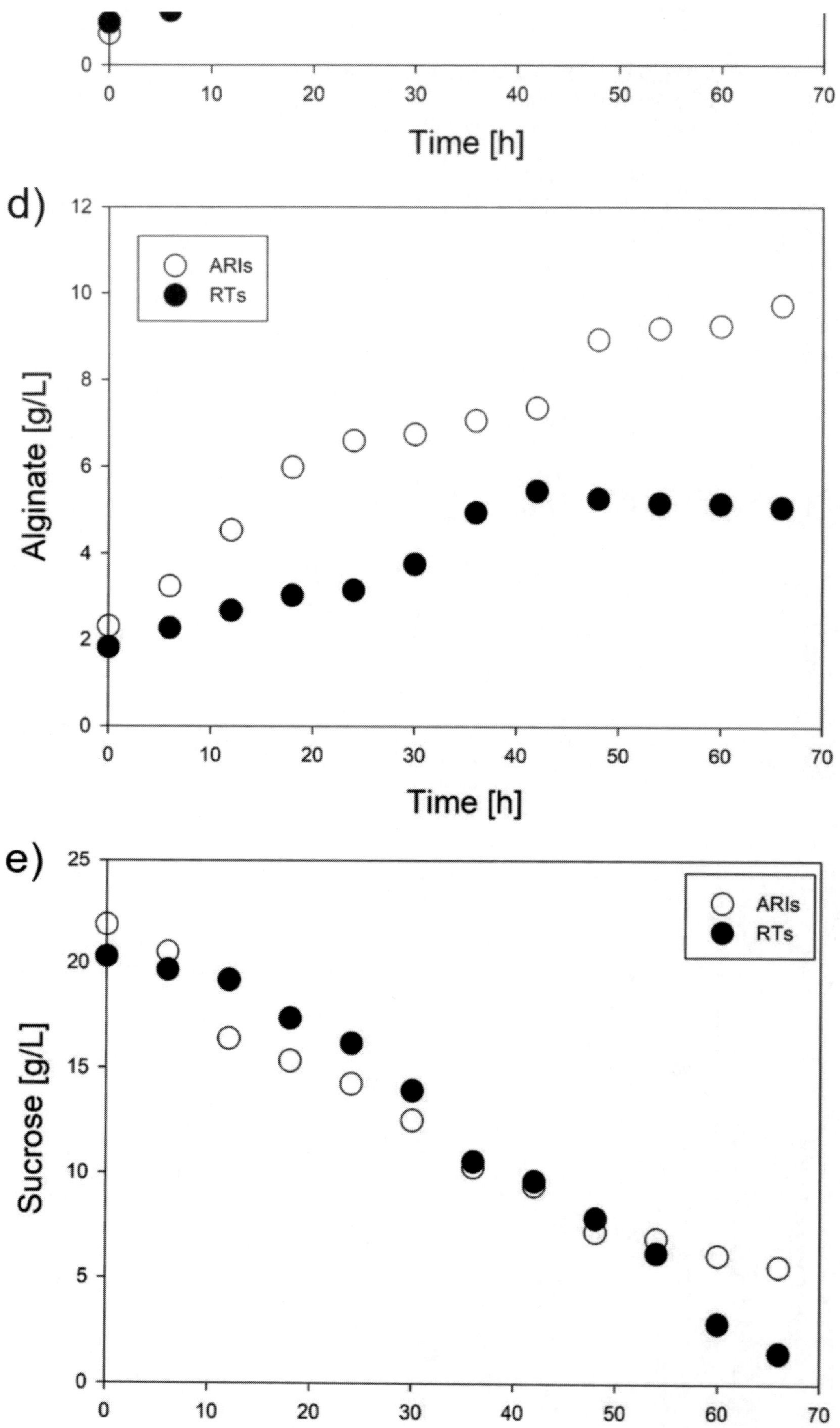
0
0
10
20
30
40
50
60
70
Time [h]
d)
Alginate [g/L]
0
2
4
6
8
10
12
ARIs
RTs
0
10
20
30
40
50
60
70
Time [h]
e)
Sucrose [g/L]
0
5
10
15
20
25
ARIs
RTs
0
10
20
30
40
50
60
70

Time [h]

735s Optimizing Energy Production from Hydrothermal Gasification of Blackwater

Aristidis Mihalos, Luke Walker, Alex Paulsen, Andrew Wagner

Mainstream Engineering has developed and demonstrated production of energy from a blackwater treatment system that concentrates waste solids in the blackwater and performs subsequent hydrothermal gasification and wastewater treatment. The deployable system is designed to be self-sustaining and operate in remote, deployed locations, where simple operation under optimal conditions of a varying feedstock is critical. Unlike conventional wastewater treatment systems that operate on pre-digested sewage sludge, no pre-retention is allowed, requiring a wetter, and more raw feedstock. In addition, the energy produced must be balanced against the requirement complete removal of organic carbon from the remaining water waste to meet stringent disposal requirements. To meet these conditions, improved understanding of continuous flow gasification of these feedstocks is needed. Mainstream conducted experimental investigation of varying blackwater feedstocks (solids loading 1 – 6%) in batch (15 – 60 minutes) and continuous (1 – 6 minutes) under a range of temperatures and catalysts and resulting gas composition and yield determined. The resulting mechanisms, impact of feedstock on yield, and best conditions for producing energy were determined, indicating the ideal product breakdown resulted in low hydrogen yields and high methane yields for easy energy generation by standard process equipment.

735t Optimizing Lithium Carbonate Crystallization Via a Novel Sieve Technology

Mohammad J Seyed Sabour, Bahman Ghorashi

Despite its historical significance as a method for separation and purification, crystallization is still a challenging process to optimize. Given that crystallization is a pivotal technique in mineral extraction from effluent; and lithium salts, such as lithium carbonate, stand out as key components extractable from rejected water in desalination plants, this study centers on optimizing crystallization for lithium carbonate using innovative sieve plates. This approach underscores the process's efficiency and its crucial role in sustainable water management and mineral recovery. This ongoing research involves the design and evaluation of a novel sieve plate for the crystallization of Lithium Carbonate in two distinct setups: a single-unit Li_2CO_3 solution and a multi-stage unit flow-through crystallizer. However, the focus of this study is on the behavior of a single-unit crystallizer. The study examines how the extent of sieve plate coverage and surface texture can influence the yield and granularity of the Lithium salt crystals in a single unit, and examines the impact of sieve plate exposure to several different solution-air interfaces.

We anticipate the use of sieve plates with about 60% coverage and enhanced exposure to solution-air interfaces will lead to improved salt yields. Additionally, we explore innovative alternative materials for the construction of sieve plates that could favorably influence nucleation rates and crystal growth.

By diving deeper into the geometry and surface roughness of the sieve plates, we explore variations in pore size, shape, and distribution. We submit that the geometry can significantly impact the fluid dynamics around the crystals, potentially leading to a more uniform crystal size and faster growth rate. Furthermore, we investigate the effects of enhanced system control including temperature, optimum flow rate, supersaturation levels, and solution purity to determine if a more precise controlled environment can lead to enhanced reproducibility and optimization of crystal size and quality. It is further anticipated that alternative approaches, such as pulsating flows or varying flow patterns could reveal new dynamics in crystal growth.

735u Contextualising Urban Sanitation Solutions through Complex Systems Thinking: A Case Study of the South African Sanitation System

Craig Sheridan, Andrew Thatcher, Tracy-Lynn Field, Diane Hildebrandt, Michael Kidd, Sandrama Nadan, Leslie Petrik, James Topkin

South Africa's urban sanitation infrastructure is unfit-for-purpose and collapsing, leading to significant environmental and health risks. Many alternative sanitation solutions exist, but their implementation has not produced successful roll-out to scale in this context. These failures are likely due to many interactive, systemic factors. This research adopts a transdisciplinary knowledge synthesis approach to map the possible systemic factors to provide engineers and policymakers with relevant information to identify contextual risk factors for potential failure and emerging problems and novel approaches. Building from the narrative literature reviews and the joint fact-finding approach, the sustainable systems-of-systems framework is used as the analysis tool to identify important elements of the complex systemic context. The output is a set of nine interweaved contextual constraints. The value of the contextual constraints is illustrated by application to three commonly implemented sanitation solutions (chemical toilets, urine-diverting dry toilets, and decentralised wastewater treatment systems) where relevant contextual components of the success (or failure) of the solutions are highlighted. The framework is not intended to replace existing linear-thinking decision-making tools, but to provide important contextual information to augment existing decision-making.

735v Integrated Combinatorial Synthesis,Characterization, and Test Platform for Lithium-Ionbattery Cathode Materials

Yunhao Xiao

An integrated combinatorial synthesis, characterization, and testing methodology and platform is proposed and developed for high-throughput exploration of Li-ion rechargeable battery cathode chemistries. This article describes the design of the platform's combinatorial synthesis part and its prototype performance. A key design element is a multi-gear powertrain with a unique milling-force adjustable pestle that ensures high compatibility with varying types of electrode materials. To demonstrate the prototype, $LiNiO_2$ was prepared via a solid-state route under various operating conditions. We evaluated the effects of processing parameters, including milling force and mixing speed, on the synthesized electrode's physical properties and electrochemical behavior. The synthesized $LiNiO_2$ materials show excellent electrochemical performance comparable to manually synthesized $LiNiO_2$.

735w Molecular Dynamics Study of Methane Hydrate Dissociation Under Silica Confinement: Role of Salinity

Bhavesh Moorjani, Jhumpa Adhikari, Samik Hait

In nature, CH_4 hydrate reservoirs are found in the seafloor and permafrost regions. In most of the reservoirs, gas hydrates are present in pores of sedimentary rocks which are of silica and carbonate types. Bagherzadeh *et al.* in their work found that the bulk phase dissociation of CH_4 hydrate is faster compared to dissociation under silica confinement. They also showed that the presence of a bound layer near the silica surface limits the interactions between the surface hydroxyl group and hydrate water molecules so an intermediate dissociation rate is obtained in that case. [1] Simulations performed by Fang *et al.* showed that the silica layer provides adsorption sites to CH_4 nanobubble which limits the migration of nanobubble from pore to bulk phase. [2] This indicates the deviation of hydrate properties under confinement compared to the bulk phase and hence confinement will also affect the recovery process. Thus, it is of vital importance to strengthen our understanding of gas hydrate behaviour under confinement to represent actual reservoir scenarios and improve the efficiency of methane extraction from hydrate reservoirs. Towards this goal, we have performed molecular dynamics (MD) simulations of the CH_4 hydrate phase in contact with liquid water which is placed under the confinement of silica walls. We have considered three systems: (1) liquid water – hydrate (Lw-H) bulk phases (2) Lw-H phases under silica confinement (3) Lw-H phases under confinement with NaCl added in aqueous phase. The number of NaCl atoms in the system corresponds to the ~3.5 wt % concentration of NaCl in water which is near to the seawater conditions. [3] Unlike the previous confinement simulations reported in the literature [1,2,3] silica atoms in our simulations are neither position restrained nor kept frozen during the MD run; we have allowed silica atoms to move freely according to their natural dynamics based on interactions calculated from selected forcefields. Simulations have been performed in the constant volume and temperature (NVT) ensemble at 290, 300, 305, and 310 K temperatures. The choice of these temperatures is based on our findings from the bulk Lw-H phase melting point of CH_4 gas hydrate for a combination of TIP4P/2005 model of water and TraPPE-EH model. Through these simulations, we want to study mainly four aspects which are: (1) confinement effect on methane hydrate dissociation, (2) effect of temperature on hydrate dissociation under confinement, (3) effect of salinity, i.e., the addition of NaCl to the liquid phase on methane hydrate dissociation, and (4) nanobubble interaction with the silica surface. On comparing these results to Lw-H phases under confinement, we find that when Lw-H phases are confined in silica slab, melting point suppression occurs. From our bulk phase simulations, we observe that the methane hydrate dissociates at 310 K temperature while the hydrate phase remains stable at other temperatures i.e., 290, 300, and 305 K. Apart from that, rate of hydrate dissociation at 310 K is much faster under confinement; such faster dissociation under confinement can be explained by the interaction of surface -OH groups of silica slab with hydrate water molecules via the hydration shell which affects the hydrogen bonding of hydrate structure near the silica slab. From the contour plots of the F_3 order parameter, we observe hydrate dissociation occurring in a shrinking core manner [1] (refer to Figure-1). We observe the formation of methane nanobubble (either cylindrical or spherical in shape) in our simulations. Interestingly we notice that regardless of the shape of the nanobubble, the nanobubble formed in all cases tends to go towards the slab substrate and stick to the surface. A similar observation is also reported by some of the previous confinement simulation work. [2] Also, we notice the formation of an ordered bound water layer on the hydrophilic silica slab under the methane nanobubble. Water molecules in this bound water layer are ordered in square lattice form whereas in other parts of the silica surface there is no specific ordering of water molecules. This raises an interesting question: "Does the tendency of the nanobubble to adhere to the silica surface affect the recovery of methane?" Answering these questions will help to enhance our understanding of the overall methane recovery process at reservoir conditions.

References:

[1] Bagherzadeh, S.A., Englezos, P., Alavi, S. and Ripmeester, J.A., 2012. Molecular modeling of the

dissociation of methane hydrate in contact with a silica surface. *The Journal of Physical Chemistry B, 116*(10), pp.3188-3197.

[2] Fang, B., Ning, F., Ou, W., Wang, D., Zhang, Z., Yu, Y., Lu, H., Wu, J. and Vlugt, T.J., 2019. The dynamic behavior of gas hydrate dissociation by heating in tight sandy reservoirs: A molecular dynamics simulation study. *Fuel, 258*, p.116106.

[3] B. Fang, T. Lü, L. Cheng, D. Wang, Y. Ni, B. Fan, J. Meng, T.J.H. Vlugt, and F. Ning, "Negative Effects of Inorganic Salt Invasion on the Dissociation Kinetics of Silica-Confined Gas Hydrate via Thermal Stimulation," Energy Fuels 36(12), 6216–6228 (2022)

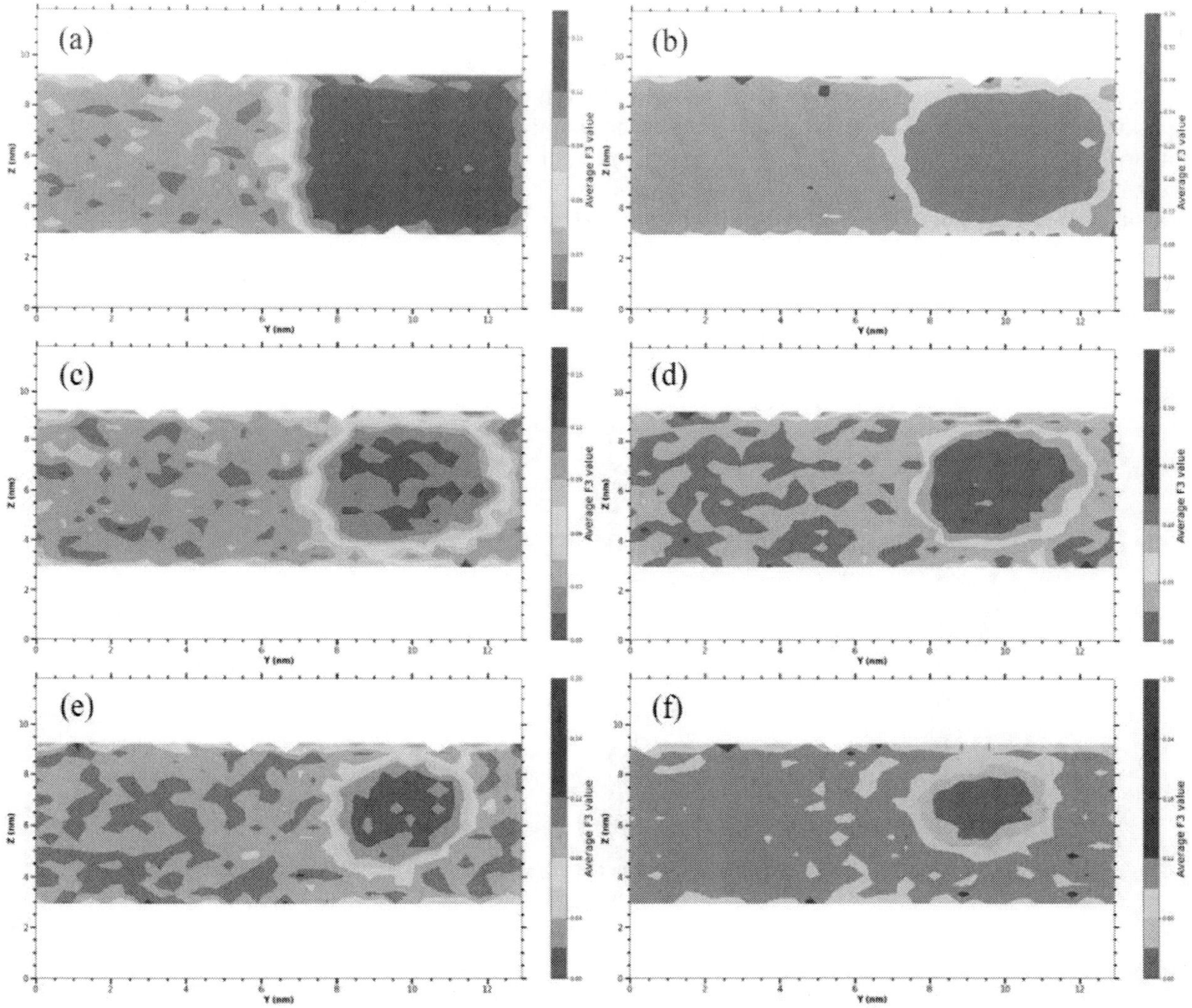

Figure-1: Contour plot of F_3 order parameter in ZY plane for 310 K run at (a) 0 ns, (b) 1 ns, (c) 2 ns, (d) 3 ns, (e) 4 ns, and (f) 5 ns (blue colour indicates water region and red colour indicates hydrate region)

735x Inferring the Shape of an Object inside of a Draining Tank Via Bayesian Statistical Inversion

Gbenga Fabusola, Cory Simon

In an inverse problem of reconstruction, we wish to predict the input to a physical system that produced some measured output—i.e., predict the cause of an observed effect. Bayesian statistical inversion is an approach to inverse problems that allows us to (i) incorporate prior information about the input into the solution and (ii) quantify uncertainty about the unknown input.

We tackle the following reconstruction problem with Bayesian statistical inversion. We possess an opaque, open-top tank with known geometry. A heavy, solid object is inside the tank whose geometry is unknown. To gather information about the geometry of this object via a "liquid level scan", we fill the tank with liquid then drill a hole in its side, allowing the liquid to drain autonomously. From measurements of the liquid level in the tank over time, we wish to infer the shape of the object inside the tank—its area as a function of height. We employ Bayesian statistical inversion so we can (i) mitigate spurious "choppiness" in the inferred area function, caused by measurement noise, by imposing a smoothness prior and (ii) quantify uncertainty about the shape of the object inside the tank.

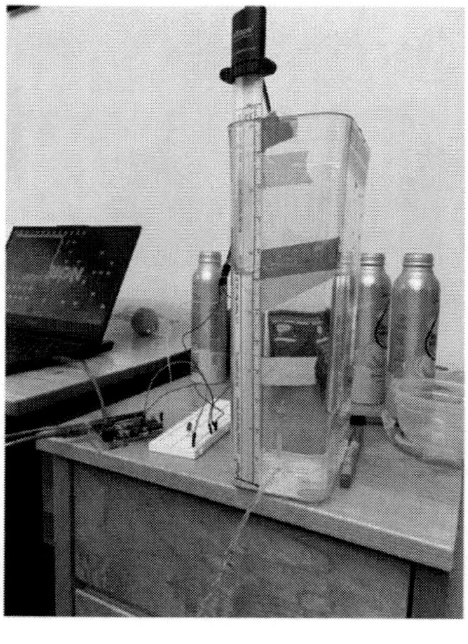

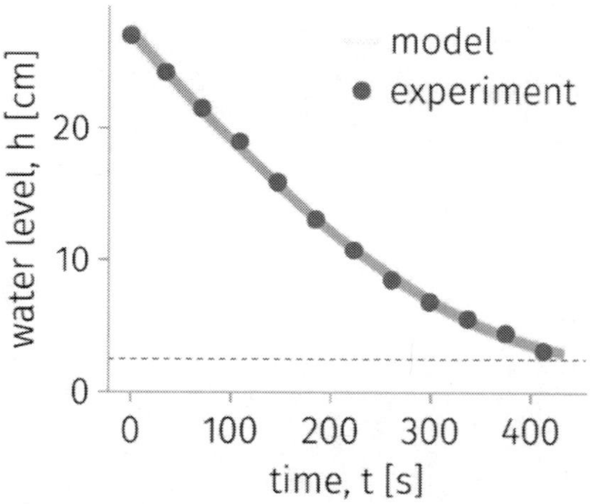

735y Stickiness Measurements in Relation to Particle Process Setting

Gabriel Meesters, Aleksander van Zanten, Jan de Jonge

Measuring the stickiness of powders in relation to their temperature and moisture content is of significant importance in the production of powders, esp. from liquids. Specifically in spray dryers, fluid bed and drum dryers, where heat is used to remove the water. Also in coating related processes, the stickiness gives valuable information on spray rates of the coating solution in relation to the maximum temperature and humidity in the coater that can be set to prevent the particles from sticking to each other. In this talk some known (DVS, DSC, Anton Paar, etc) and some novel techniques (e.g. Granudrum) will be discussed and the limitations of these techniques will be given.

735z Unraveling Dynamic Surface Restructuring and Intricate Reaction Networks in Catalysis Via Integrated DFT-Microkinetic Modeling

Kunran Yang, Bo Yang

Computational catalysis has emerged as a powerful tool for rational catalyst design, particularly in heterogeneous catalysis, offering cost-effective and predictive capabilities. However, accurately modeling realistic and dynamic catalytic conditions encountered in practical systems remains a challenge. Conventional computational techniques often struggle to capture the surface restructuring induced by temperature, pressure, and gas compositions, as well as the intricate reaction network complexity. To address these limitations, we have developed an integrated approach combining density functional theory (DFT) and microkinetic simulations to study catalytic processes under realistic conditions, with a focus on complex reaction networks.

In this talk, I first illustrate the surface restructuring of Cu-based single-atom alloy (SAA) catalysts under reaction conditions. Our findings reveal the thermodynamic instability of RhCu-SAA and NiCu-SAA structures upon acetylene adsorption, leading to the aggregation of Rh and Ni atoms on the surfaces. Conversely, the PdCu-SAA and PtCu-SAA surfaces favor the selective hydrogenation of acetylene to ethylene, consistent with experimental observations. Additionally, we probed the surface aggregation of PtCu-SAA during butadiene hydrogenation reactions, where butadiene adsorption induces the aggregation of surface Pt atoms. Microkinetic analysis successfully explained the experimental butene isomer product distribution, a feat that DFT calculations alone could not achieve.

The second topic focuses on the selective catalytic oxidation of NH_3 (SCO-NH_3) on Ag catalysts. While Ag exhibits good reactivity and N_2 selectivity, the mechanisms, detailed kinetics, and the relationship between surface structure and reactivity/selectivity under oxidizing environments remain unclear. Our studies on Ag(111) and Ag(211) surfaces revealed the preference for either O-assisted or OH-assisted oxidation pathways, depending on the intermediates and reaction temperatures. Microkinetic results predicted the formation of O-covered Ag surfaces under realistic SCO-NH3 conditions on pure Ag catalysts. We systematically investigated SCO-NH_3 mechanisms on surface-oxidized Ag catalysts. Notably, these oxidized Ag surfaces exhibited intermediate activity compared to Ag(111) and Ag(211), along with strong oxygen storage abilities and reactivity from both adsorbed surface oxygen and lattice oxygen species. These insights highlight the importance of considering Ag catalyst oxidation during SCO-NH_3 and facilitate the design of novel SCO-NH_3 catalysts.

Overall, the integration of DFT and kinetic calculations accelerates catalyst design and provides molecular-level understanding of reactions and catalyst structural evolutions under various conditions, offering a powerful approach to unravel dynamic surface restructuring and intricate reaction networks in catalysis.

References

K. Yang, B. Yang, Identification of reaction mechanisms and role of surface oxygen in selective catalytic oxidation of ammonia over surface-oxidized Ag surface: A DFT and microkinetic study, In preparation

K. Yang, B. Yang, Identifying the reaction network complexity and structure sensitivity of selective catalytic oxidation of ammonia over Ag surfaces. ***Applied Surface Science*** 2022, *584*, 152584, doi.org/10.1016/j.apsusc.2022.152584

K. Yang, B. Yang, Identification of the Active and Selective Sites over a Single Pt Atom-Alloyed Cu Catalyst for the Hydrogenation of 1,3-Butadiene: A Combined DFT and Microkinetic Modeling Study. ***Journal of Physical Chemistry C*** 2018, *122* (20), 10883-10891, doi.org/10.1021/acs.jpcc.8b01980

K. Yang, B. Yang, Surface Restructuring of Cu-based Single-atom Alloy Catalysts under Reaction Conditions: The Essential Role of Adsorbates. ***Physical Chemistry Chemical Physics*** 2017, *19* (27), 18010-18017, doi.org/10.1039/C7CP02152F

735ab Molecular Simulation Methods to Probe Dynamics and Thermodynamics in Nucleoprotein Systems

Lev Levintov

Research interests: The focus of my past research has been the computational modeling and characterization of various biomolecular systems, including nucleic acids, proteins, and membrane systems. As part of my future work, I am interested in expanding my research work on RNA molecules by characterizing the energetic and structural effects of chemical modifications in DNA molecules by performing alchemical free energy calculations of these modifications. Results from these simulations will be useful in the design and functionalization of DNA nanostructures which can be further used in the drug delivery process. Furthermore, I am interested in applying computational modeling in understanding how the host-cell DNA strands are properly positioned prior to integration of the viral DNA genome in the viral intasome complex as well as the subsequent conformational changes which the intasome undergoes during the multistep integration process of the viral genome into the host-cell genome. Insights from these simulations will be useful in designing novel integrase strand transfer inhibitors (INSTIs). As my last project, I would like to develop coarse grained models of DNA molecules to characterize DNA packaging into chromatin fibers which is an essential biophysical process in cells.

PhD Dissertation: *Molecular Simulation Studies of Dynamics and Interactions in Nucleic Acids.* Advisor: Harish Vashisth at University of New Hampshire.

Postdoctoral Projects: *Characterization of the Molecular Architecture of the HIV or Related Viruses*. Under the supervision of Prof. Juan R. Perilla.

Studying Transmembrane Protein Dimerization Using Coarse Grained Simulations. Under the supervision of Prof. Harish Vashisth.

Successful Proposals: Juan R. Perilla Postdoctoral Researcher. Assisted in writing and gathering preliminary data for ACCESS and DARWIN supercomputing facilities.

Additional Proposal Writing Experience: Harish Vashisth Postdoctoral Researcher. Assisted in writing and gathering preliminary data for the NSF MFB project.

Teaching interests: During my doctoral and postdoctoral assignments, I had the opportunity of training several undergraduate and graduate students in developing skills and knowledge necessary to conduct molecular simulations. As a TA, I had the opportunity to teach both core chemical engineering courses such as unit operations and transport phenomena as well as an elective course to students with diverse backgrounds. As a chemical engineering faculty, I am ready to teach any core chemical engineering course or any other elective that fits my professional background. As an example of an elective course, I would like to introduce students to the fundamentals of bioengineering/biophysics while further emphasizing how molecular modeling, simulation, and visualization techniques can assist us in understanding biomolecular structures.

Research experience: During my doctoral research I focused on studying dynamics and interactions in viral nucleic acids. Specifically, I have worked on the HIV-1 RNA molecules and their interactions with ligands of various sizes, ranging between small inhibitory molecules and peptides with inhibitory properties. Furthermore, I have utilized a toolkit of enhanced sampling and path sampling techniques to characterize several essential biophysical processes in nucleic acids: ligand (un)binding and base flipping mechanism. These projects provide additional understanding of how local and global conformational transitions in RNA facilitate interactions of RNA with a diverse set of ligand molecules. My postdoctoral experience has further expanded the scope of systems and methodology which I am proficient in. Specifically, I worked on the retroviral replication process with the focus on studying how clinically-approved drugs perturb essential steps in the replication cycle of HIV-1. During that period, I collaborated with experimental groups to refine experimentally resolved structures and fill in the gaps that are missing from experimental data. I also gained experience in writing first-authored manuscripts, co-authored manuscripts while providing guidance to a PhD student, as well as working in large collaborative projects exchanging data with researchers from across and outside the country.

Selected Publications:

Levintov, L., Paul, S., and Vashisth, H. (2021) "Reaction coordinate and thermodynamics of base flipping in RNA." ***J. Chem. Theory Comput.***, 17 (3), 1914-1921.

Singer, M. R., Dinh, T., **Levintov, L.**, Annamalai, A. S., Rey, J. S., Briganti, L., Cook, N. J., Pye, V. E., Taylor, I. A., Kim, K., Engelman, A. N., Kim, B., Perilla, J. R., Kvaratskhelia, M., and Cherepanov, P. (2023). "The mechanism of HIV-1 integrase aggregation by allosteric inhibitors." ***mBio.***, 14 (1), e03560-22.

Krebs, A. S., Liu, H. F., Zhou, Y., Lancheros, J. S. R, **Levintov, L.**, Shen, J., Howe, A., Perilla, J. R., Bartesaghi, A., Zhang, P. (2023). "Molecular architecture and conservation of an immature human endogenous retrovirus." ***Nat. Commun.***, 14 (1), 5149.

Levintov, L., and Vashisth, H. (2024). "Structural and Computational Studies of HIV-1 RNA." ***RNA Biol.***, 21 (1), 1-32.

Levintov, L., and Vashisth, H. (2024). "Adenine Methylation Enhances the Conformational Flexibility of an RNA Hairpin Tetraloop." ***J. Phys. Chem. B***, 128 (13), 3157–3166.

735ac Real-Time Irrigation Scheduling in Precision Agriculture: Comparison of Simulation-Based Optimization Approaches

Jisung Jang, Tian Di, Q. Peter He

This work investigates the application of simulation-based optimization methods within the Decision Support System for Agrotechnology Transfer (DSSAT) framework, aiming to enhance sustainability of agriculture against the global climate change and water scarcity. DSSAT is a software application program that comprises crop simulation models for over 42 crops, which has gained wide acceptance and has been used to simulate various applications at different spatial and temporal scales (Jones et al. 2003). This work explores a computationally efficient direct search method in the form of pattern search and a computationally more expensive population-based method of multi-objective genetic algorithm (MOGA). The objective is to optimize real-time irrigation scheduling (i.e., irrigation time/date and amount) to maximize crop yield as well as water use efficiency (WUE) under variable weather conditions.

Pattern search is a direct search method used for optimization problems where the gradient is not available (Torczon.1997). In the context of DSSAT, pattern search is a good choice since the derivative information is hard to obtain. In addition, pattern search is simple and flexible for efficient exploration of high-dimensional spaces and handling non-smooth or even non-continuous DSSAT objectives. However, pattern search employs a local search strategy, which may have difficulty avoiding local optima. On the other hand, MOGA utilizes a population-based search strategy by maintaining a population of diverse candidate solutions and iteratively applies genetic operators (selection, crossover, mutation) to generate new solutions. The objective function is evaluated for each individual in the population, guiding the selection of solutions for the next generation. This population-based approach allows MOGA to explore a wider search space and potentially avoid local optima. Another advantage of MOGA is its explicit handling of multiple (conflicting) objectives (Deb et al., 2002).

In this study, we simulate the growth and water usage of maize (corn) in the Piracicaba region of Brazil, referencing soil properties, and management practices within the DSSAT system. We utilize 36 years of weather data, spanning from 1985 to 2020, sourced from NASA's POWER (Prediction Of Worldwide Energy Resources) project. We compare the performances of pattern search and MOGA to the performance of the DSSAT built-in irrigation management strategy in terms of crop yield and WUE. We also investigate their flexibility in shifting weight between yield and WUE to address different conditions (e.g., water sufficient vs. water scarce regions). Finally, we compare the performances of the above-mentioned methods under different levels of weather forecast uncertainties.

Reference

- Jones, James W., et al. 'The DSSAT cropping system model.' *European journal of agronomy*3-4 (2003): 235-265.
- Torczon, Virginia. 'On the convergence of pattern search algorithms.' *SIAM Journal on optimization*1 (1997): 1-25.
- Deb, Kalyanmoy, et al. 'A fast and elitist multiobjective genetic algorithm: NSGA-II.' *IEEE transactions on evolutionary computation*2 (2002): 182-197.

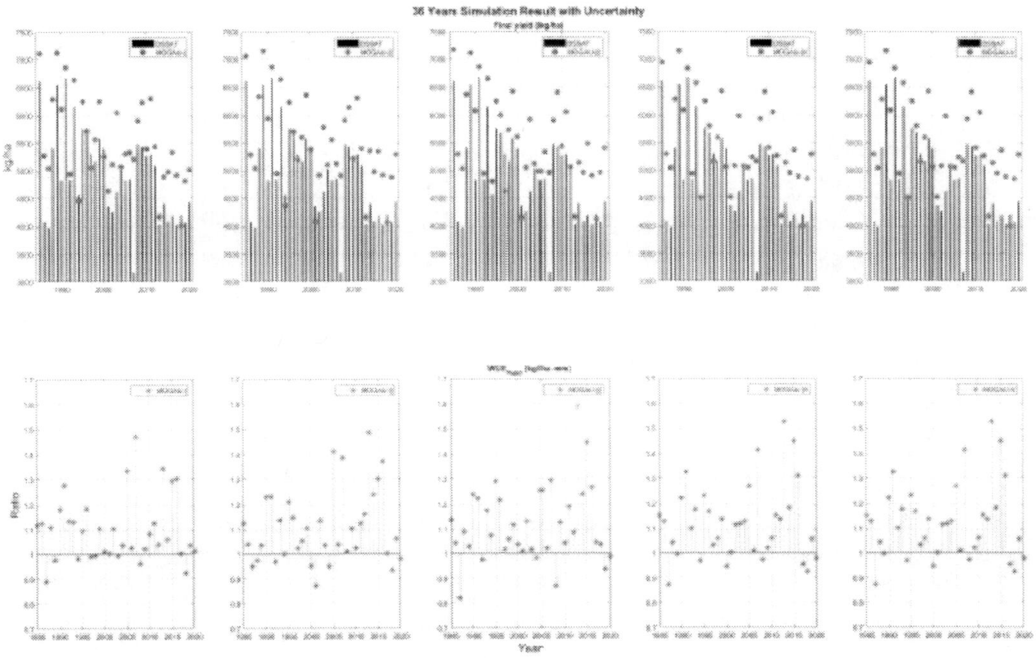

Figure 3.10: Result of MOGA with uncertainty from 1985 - 2020

Table 3.6: Average With Uncertainty: Performance Evaluation (36 Years)

Method	Yield (kg/ha)				Irrigation (mm)			$WUE_{Y/I}$ (kg/ha·mm)		
	Final	Diff.	Std	Ratio	Total	Diff.	Std	WUE	Std	Ratio
DSSAT	4931.75	0.00	812.27	0.00%	229.58	0.00	53.65	20.39	5.69	0.00%
MOGA-U1	5499.14	567.39	765.29	12.65%	263.67	34.08	66.39	22.35	6.93	9.61%
MOGA-U2	5487.72	555.97	780.99	12.30%	261.03	31.44	66.58	22.50	6.75	10.60%
MOGA-U3	5513.33	581.58	790.96	12.86%	260.94	31.36	67.84	22.67	7.01	11.38%
MOGA-U4	5494.25	562.50	759.84	12.58%	258.97	29.39	69.14	22.92	7.38	12.16%
MOGA-U5	5458.58	526.83	754.18	11.82%	251.78	22.19	59.32	23.03	6.80	13.43%

$WUE_{Y/I} = \frac{\text{Yield}}{\text{Irrigation}}$
Diff. – DSSAT-MOGA
Std – Standard Deviation

735ad Deciphering the Sequence-Dependent Relations between Single-Chain Protein Conformations and Their Condensed Phase Material Properties

Dinesh Sundaravadivelu Devarajan, Jeetain Mittal

The regular physiological functioning of many biomolecular condensates depends on their structure, dynamics, and rheology, imparted primarily by intrinsically disordered proteins (IDPs). Despite the significant effort dedicated to understanding how the sequence of IDPs drives their phase separation to form condensates, little is known about the sequence determinants of their single-chain dynamics and condensate material properties. In this work, we computationally uncover these relations for charged model and naturally occurring IDPs, prevalent in biocondensates. Specifically, we investigate how the arrangement of charged residues within these sequences influences their single-chain properties and condensate biophysical properties. At the single-chain level,[1, 2] we find that the chain-level dynamics correlate with changes in conformation with increasing charge segregation. Importantly, the chain-level and segmental dynamics conform to simple homopolymer models only for uniformly charge-patterned sequences of the model and natural proteins. Within the condensates,[3] we find diffusion, viscosity, and interfacial tension to change significantly with increasing charge segregation, but the changes are surprisingly similar between the model and natural proteins, despite their very different sequence compositions. Our simulations using implicit and explicit solvent models, reveal that molecular contacts within the condensates are highly similar to those within a single chain for all sequences. Remarkably, the condensate material properties of charged disordered proteins are strongly correlated with their dense phase contact dynamics and single-chain structural properties. Two important conclusions emerge from our work: (1) the need for caution in inferring IDP properties in dilute conditions using homopolymer scaling laws as is often done experimentally and (2) the potential utility of single-chain protein simulations in rapidly exploring varied protein sequences, thus expediting the design and prototyping of biopolymers with desired material properties in future.

References

(1) Sundaravadivelu Devarajan, D.; Rekhi, S.; Nikoubashman, A.; Kim, Y. C.; Howard, M. P.; Mittal, J. Effect of Charge Distribution on the Dynamics of Polyampholytic Disordered Proteins. *Macromolecules* **2022**, *55* (20), 8987-8997.

(2) Wang, J.; Sundaravadivelu Devarajan, D.; Kim, Y. C.; Nikoubashman, A.; Mittal, J. Sequence-Dependent Conformational Transitions of Disordered Proteins During Condensation. *bioRxiv* **2024**, DOI: 10.1101/2024.01.11.575294.

(3) Sundaravadivelu Devarajan, D.; Wang, J.; Szała-Mendyk, B.; Rekhi, S.; Nikoubashman, A.; Kim, Y. C.; Mittal, J. Sequence-dependent material properties of biomolecular condensates and their relation to dilute phase conformations. *Nature Communications* **2024**, *15* (1), 1912.

735ae Accelerating Simulation of Materials with Machine Learning

Ni Zhan

A new generation of advanced materials will be required for energy sustainability and quantum technology, and computation aids materials search and investigation. Molecular dynamics (MD) simulation, in particular, allows us to calculate transport properties and investigate dynamic behavior, which are difficult to probe experimentally. Challenges of MD simulation include the trade-off between accuracy and computational expense, analyzing simulation data, and uncertainty in calculated properties. My research uses machine learning to address these challenges. The trade-off between accuracy and computational expense motivates the ML potential, commonly parameterized by neural networks (NN), because they can reach DFT level accuracy at orders of magnitude faster. One system of interest is liquid alloys, and I previously trained NN potentials for liquid Ni-Al-W and integrated them with MD simulations. The training process required an iterative build-up of the training dataset with human feedback, because NNs commonly extrapolate outside their training regime, motivating an uncertainty quantification method. We develop and show that the delta method for NN potentials identifies data-points where the model extrapolates. We also describe different sources of uncertainty and difficulty of obtaining uncertainties of physical properties, especially derivatives and equilibrium properties. We obtain uncertainties of equilibrium lattice constants and bulk moduli, and show that Frequentist and Bayesian regression uncertainty are specific to model choice while Gaussian process uncertainty is model general/agnostic. My work will allow robust, automated investigation of dynamics properties at scale.

735af Advancing Separation Systems of High Brine Concentrations through Refined Thermodynamics and Process Intensification

Nazia Aslam, Wajeha Tauqir, Matthew Stuber, George Bollas

Modeling separation processes for solutions with high brine concentrations requires accurate models of separation, which inherently incorporate heat and mass transport phenomena. These models are widely utilized in the water industry. Additionally, modeling of the involved unit operations necessitates accurate thermodynamic models to compute various electrolyte thermodynamic properties, such as the mean ionic activity coefficients in concentrated aqueous multi-electrolyte solutions. However, many current models incorporate the conventional electrolyte-nonrandom two-liquid (eNRTL) [1] model that lacks accuracy at high temperatures and concentrations.

In this study, the refined-eNRTL [2] model is used to refine the implementation of thermodynamic property models in the modeling of highly concentrated brine separation systems. In prior work, Stuber et al. [3] developed models of the unit operations of concentrator systems, representing an advanced separation process based on solar-powered multi-effect distillation (MED) with an absorption heat pump (AHP). The MED system was designed with the water-energy nexus in mind, which offers improved thermal efficiency from the AHP, and reduces the dependence on water-intensive power sources by utilizing solar thermal power as its primary energy source. This system's set of unit operations served as the foundational model for this work, extending it with the inclusion of the refined-eNRTL model. The new model was developed in Pyomo with modular libraries of IDAES [1] and WaterTAP [4] to combine electrolyte thermodynamics, process modeling, simulation, and optimization for the process intensification of high brine concentration and separation systems. When accounting for non-ideality in the system, the inclusion of the refined e-NRTL model resulted in differences in mean ionic activities and a reduction in energy consumption compared to less accurate thermodynamic property methods. The system model was then used to study designs and operating conditions that intensify and optimize the process under varying amounts of water recovery in the separation process. In conclusion, this study shows how the accurate prediction of brine properties can support process intensification within the water-energy nexus, allowing for greater utilization and reuse of non-traditional water sources.

Acknowledgments

This study was supported by the National Alliance for Water Innovation (NAWI), funded by the U.S. Department of Energy, Energy Efficiency and Renewable Energy Office, Advanced Manufacturing Office.

References

[1] Lee, Andrew, Jaffer H. Ghouse, John C. Eslick, Carl D. Laird, John D. Siirola, Miguel A. Zamarripa, Dan Gunter et al. 'The IDAES process modeling framework and model library—Flexibility for process simulation and optimization.' Journal of Advanced Manufacturing and Processing 3, no. 3 (2021): e10095.

[2] Bollas GM et al. Refined Electrolyte-NRTL Model: Activity Coefficient Expressions for Application to Multi-Electrolyte Systems. AIChE J. 2008;54(6):1608-1624

[3] Stuber MD et al. Pilot demonstration of concentrated solar-powered desalination of subsurface

agricultural drainage water and other brackish groundwater sources. Desalination. 2015;355:186-196.

[4] WaterTAP contributors. WaterTAP: An open-source water treatment model library. Version 0.6. Sponsored by California Energy Commission, National Alliance for Water Innovation, and USDOE. Available at https://github.com/watertap-org/watertap

735ag Zero-Dimensional Pores in Graphene for Ion-Ion Separation

Kumar Varoon Agrawal

Strong confinement of solute inside zero-dimensional pores of graphene can differentiate ions based on their ability to partition from the bulk to the pore. Graphene film is an ideal selective layer for this if high-precision pores can be incorporated. However, precise pore size control in Å-regime has proven challenging for ion-ion differentiation due to lack of methods to finely tune pore size in this regime. Further, it is extremely challenging to avoid larger non-selective pores at the tail-end of the pore size distribution which reduces ion-ion selectivity.

Herein, we report a novel Å-scale pore size tuning tool, which incorporates a high density of ion-ion selective pores between 3.5 and 8.5 Å while minimizes the nonselective pores above 10 Å. These pores impose a strong confinement for ions, which results in extremely high selectivity from centimeter-scale porous graphene between monovalent and bivalent ions and near complete blockage of ions with hydration diameter, D_H, greater than 9.0 Å. Ion diffusion study reveals the presence of an energy barrier corresponding to partial dehydration of ions with barrier increasing with D_H. We observe a reversal of K^+/Li^+ selectivity at elevated temperature and attribute this to the relative size of dehydrated ions.

Further, we discuss a novel strategy to overcome this challenge using an electrochemical repair strategy that successfully masks larger pores in large-area graphene. 10-nm-thick electropolymerized conjugated microporous polymer (CMP) layer is successfully deposited on graphene, thanks to a strong π-π interaction in these two materials. While the CMP layer itself is not selective, it effectively masks graphene pores, leading to an unprecedented Li^+/Mg^{2+} selectivity from zero-dimensional pores with a high Li^+ ion permeation rate surpassing the performance of reported materials for ion-ion separation.

These results underscore the promise of porous two-dimensional materials for solute-solute separation when Å-scale pores can be incorporated in a precise manner. The scalable repair strategy enables the fabrication of monolayer graphene membranes with customizable pore sizes, limiting the contribution of nonselective pores, and offering graphene membranes a versatile platform for a broad spectrum of challenging separations.

735ai Direct Scale-up of a Dry Granulation Process Using a Material Lean Approach at Lab-Scale

Artur Saramago

Delivering uniform dosage forms to patients may be challenging for poorly flowing formulations through direct compression. Flowability issues and powder properties variability that spray dried amorphous solid dispersions (ASD) may bring [1], ultimately lead to higher weight variations during tableting and/or inconsistent drug content uniformity. Granulation processes (wet or dry) allow to overcome these issues through particle size and density increase [2,3]. Dry granulation is a continuous process with the advantage of being able to work with heat or moisture sensitive drugs while reducing costs compared to wet granulation. In this process, powders are fed between two counter-rotating rollers. Pressure from the rollers compacts the powder, forming thin ribbons which are then milled into granules used as drug product intermediaries for blending with excipients prior to tableting/capsule filling.

The present work is focused on a lean, streamlined material sparing approach for laboratory development to enable a direct scale-up of blending, roller compaction (RC) and tableting steps to the manufacturing facilities. For this purpose, commonly selected polymers for spray drying to enhance API bioavailability through ASD i.e., hydroxypropyl methylcellulose acetate succinate (HPMCAS) or polyvinylpyrrolidone-vinyl acetate (PVP/VA), were blended with conventional excipients used in the manufacture of oral dosage forms (microcrystalline cellulose, fumed silica, croscarmellose sodium, and magnesium stearate, at 50% w/w ASD content).

Ribbons were manufactured in a Gerteis Mini-Pactor at different densification conditions by changing roller compaction gap and force, generating granules with different particle size distribution and density. The roller compaction conditions to obtain different ribbon at gap solid fraction were estimated by leveraging a mechanistic RC model based on Johanson's theory.[4].

At lab scale, leveraging compaction simulation capability (STYL'One Nano) and a lab scale granulator (FlexMill-Lab OsciloWitt), the roller compaction process train was simulated. To manufacture slugs at lab, the ribbon densification time at compaction simulator was estimated via roll geometry speed and nip angle simulated and the roller compactor milling tip speed was mimicked at lab. Ribbon (from manufacturing equipment) and slugs' (from laboratory equipment) SF and their resulting particle size distribution and Hausner ratio were compared. An offset was observed between ribbons at gap / out of gap and slugs in die / out of die SF. The slugs presented a narrower difference between in die and out of die due to less elastic recovery. Between polymers at the same set of RC conditions, larger offset was observed for the HPMCAS based formulation. Aligned with mechanical properties (PVP/VA often used as a binder whereas HPMCAS is a cellulose based providing higher flexibility within the polymeric chain). Nonetheless, it was observed that slugs and ribbons with matching SF out of die or gap, respectively, yielded similar particle size distribution and Hausner ratio. Additionally, the application of higher roller compaction forces and narrower gap settings resulted in granules with larger particle size, as result of increased SF of the ribbons.

The tablets processability from manufacturing and laboratory routes were compared through compression profile analysis, friability, and disintegration results. The compression profile used simulated an 8 EU-D stations Korsch XL-100 at a nominal throughput. The results confirmed that similar ribbon SF out of gap and slugs' SF out of die resulted in tablets with matching processability. Comparing the tableting results obtained from different RC conditions, an earlier overcompression onset for the trials produced at higher densification conditions (higher ribbon solid fraction) was observed.

The current work demonstrated it is possible to develop the dry granulation process using a lean methodology at the lab-scale. This methodology significantly minimizes the API needs and accelerates process development by enabling right-first-time scale-up from laboratory to industrial scale. This ultimately expands capacity, while reducing costs, materials and time which are critical in early-stage programs.

References:

[1] S. Page and R. Maurer, "Downstream Processing Considerations," in *Amorphous Solid Dispersions: Theory and Practice*, N. Shah, H. Sandhu, D. S. Choi, H. Chokshi, and A. W. Malick, Eds. Springer-Verlag New York, 2014, pp. 395–417.

[2] S. A. Howard, "Solids: Flow Properties," no. June 2016, doi: 10.1081/E-EPT3-100200005.

[3] W. J. Sun, J. Rantanen, and C. C. Sun, "Ribbon density and milling parameters that determine fines fraction in a dry granulation," *Powder Technol.*, vol. 338, pp. 162–167, 2018, doi: 10.1016/j.powtec.2018.07.009.

[4] Sousa, R., *et al.*, "Roller Compaction Scale-Up Made Simple: An Approximate Analytical Solution to Johanson's Rolling Theory," *J. Pharm. Sci.*, 2020, doi: 10.1016/j.xphs.2020.05.004.

735aj Population Balance Modeling of the Impacts of Batch Size, Flow Rate, and Imperfect Mixing during Nanomilling of Drug Suspensions

Hamidreza Heidari, Nontawat Muanpaopong, Gulenay Guner, Helen Yao, Don Clancy, Ecevit Bilgili

Enhancing bioavailability is a crucial focus in pharmaceutical research and development, dedicated to improving the therapeutic effectiveness of drug formulations. Nanoparticles have emerged as highly promising vehicles due to their large surface area, enhanced saturation solubility, sustained release, and ability for targeted delivery to improve bioavailability [1-3]. Nanomilling of drugs via wet stirred media milling (WSMM) is the most widely used process in the pharmaceutical industry to produce stable drug nanosuspensions [4]. The widespread adoption of WSMM stems from its robust, reproducible, scalable, solvent-free, and environmentally friendly operation [2,6,7].

A quantitative understanding of the impacts of process-design parameters through modeling, particularly mechanistic or first-principles-based modeling, can be highly beneficial for process development and optimization of the WSMM process [8,9]. Population balance models (PBMs) are often employed for simulation, control, and optimization of particulate processes [10,11]. In addition to describing the spatial and temporal evolution of the particle size distribution (PSD), PBMs have also been utilized to identify particle breakage mechanisms [12,13]. While several PBM studies exist for simulating batch WSMM [14] and multi-pass continuous WSMM [15], Annapragada and Adjei [16] offered the first credible PBM structure for a recirculating WSMM system. However, their model lacked thorough experimental verification and exhibited significant deviations from limited experimental data.

We aimed to develop and validate a cell-based PBM that allows for elucidating the impacts of batch size, suspension volumetric flow rate, and imperfect mixing within the mill chamber on the PSD evolution. Several experiments were conducted using a recirculating WSMM to produce nanosuspensions of a model poorly soluble drug, fenofibrate. Experimentally, the PSD evolution at various batch sizes (118 mL, 236 mL, and 512 mL) and suspension volumetric flow rates (63, 126, and 250 mL/min) was measured, where the baseline conditions had 236 mL batch size and 126 mL/min volumetric rate. In the first part of this study, four different specific breakage rate functions (Models A–D) were fitted to experimental PSD data at the baseline conditions, assuming a well-mixed mill. A full back-calculation method, employing a dynamic global optimizer and ODE solver, was used to estimate the breakage parameters and discriminate between the four models based on the sum-of-squared residuals (SSR). Among the four specific breakage rate functions examined, Model C, consisting of a product of power-law and logistic functions with a transition particles size, provided the best fit to the experimental PSD data at baseline conditions, with the lowest SSR. Model A (traditional power-law model) and Model B (a Heaviside unit function coupled with the power-law to capture the grinding limit) did not perform well. **Figure 1** displays the PBM simulation using the best-fitting breakage rate function (Model C) alongside the experimental PSD evolution over the initial 32 min and from 64 to 128 min for the baseline condition (batch size of 236 mL and flow rate of 126 mL/min), respectively. The PBM simulations aligned well with experimental observations, capturing the shift in PSD toward finer particles and the narrowing of the distribution as milling progressed. However, the PBM struggled to accurately capture the bimodal PSD observed during the initial stages of milling, potentially arising from the presence of aggregates. This limitation highlights the need for future modeling efforts to incorporate aggregation terms in the PBM to improve its predictive capabilities. The discrimination of the four models revealed the criticality of incorporating a transition particle size commensurate with the notion of a grinding limit (Models C and D); however implausible discontinuity was observed in Model B case.

In the second part of the study, the best-fitting breakage rate function (Model C) was then used to simulate the impacts of the batch size and volumetric flow rate of the suspensions. The PBM predictions were compared with respective experimental data for model validation. Both experiments and PBM simulations demonstrated that an increase in batch size necessitates a proportionate increase in milling time to maintain the same PSD. This can be achieved by keeping the effective milling time the same for different batch sizes. The effective milling time is equal to the product of number of batch turnovers (theoretical passages through the mill) and single-pass mean residence time in the mill. The suspension volumetric flow rate was found to have an insignificant impact on the PSD evolution, as the effective milling time becomes independent of the flow rate. A key outcome of this study was the demonstration that maintaining the same effective milling time is crucial for achieving consistent product quality when scaling up or changing batch sizes. This principle was verified through simulations and experiments, where the PSDs at different batch sizes were found to be identical when the effective milling time was kept constant. These findings contradict previous simulation studies [16], which suggested a notable effect of the suspension flow rate. However, the experimental results in this study support our PBM simulations, indicating that the volumetric flow rate has a minimal influence on the PSD in a recirculating WSMM system.

In the third part of the study, the potential impact of imperfect mixing in the mill chamber was theoretically examined by considering multiple cells and different back-mixing flow ratios (R) in the cell-based PBM. The number of cells (n) and R modulate the residence time distribution (RTD) behavior, accounting for deviations from perfect mixing. An intriguing finding from the PBM simulations was that the PSD was not affected by the RTD behavior, as modulated by the n and R in the cell-based PBM, after a few turnovers in the recirculation mode.

In conclusion, this study provides valuable insights into the operation of a recirculating WSMM for nanoparticle production through the development and experimental validation of a PBM with a novel specific breakage rate function. The findings highlight the criticality of maintaining the effective milling time during scale-up or batch size changes to ensure consistent product quality. The suspension volumetric flow rate was found to have a negligible impact on the PSD evolution, contrary to previous simulation studies. Furthermore, the simulations suggest that the PSD is insensitive to the RTD behavior after a few turnovers in the recirculation mode, potentially simplifying the modeling approach by eliminating the need for detailed RTD studies. Despite some limitations, such as the inability to capture bimodal PSDs arising from aggregation during the early milling times, the newly formulated PBM offers valuable guidance for process development and scale-up of recirculating WSMM operations. Future work should focus on incorporating aggregation terms in the PBM to improve its predictive capabilities, as well as expanding the model to account for the impacts of all process parameters. Overall, this study demonstrated that a novel PBM enabled us to elucidate the impacts of the batch size, suspension volumetric flow rate, and imperfect mixing and gain fundamentals insights into the recirculating WSMM operation. It is hoped that this study will guide the batch size and scale-up studies in pharmaceutical industry.

References

1. Tanaka, Y.; Inkyo, M.; Yumoto, R.; Nagai, J.; Takano, M.; Nagata, S. Nanoparticulation of probucol using wet milling to improve bioavailability. Drug Dev. Ind. Pharm. 2012, 38, 1015-1023. [https://doi.org/10.3109/03639045.2011.640299]

2. Li, M.; Azad, M.; Davé, R.; Bilgili, E. Nanomilling of drugs for bioavailability enhancement: a holistic perspective. Pharmaceutics 2016, 8, 17. [https://doi.org/10.3390/pharmaceutics8020017]

3. Malamatari, M.; Taylor, K.M.; Malamataris, S.; Douroumis, D.; Kachrimanis, K. Pharmaceutical nanocrystals by wet milling. Drug Discov. Today 2018, 23, 534-547. [https://doi.org/10.1016/j.drudis.2017.10.021]

4. Bhakay, A.; Rahman, M.; Dave, R.N.; Bilgili, E. Bioavailability enhancement via nanocomposites: Formulation–Processing aspects. Pharmaceutics 2018, 10, 86. [https://doi.org/10.3390/pharmaceutics10030086]

5. Merisko-Liversidge, E.; Liversidge, G.G.; Cooper, E.R. Nanosizing for poorly water-soluble drugs. Eur. J. Pharm. Sci. 2003, 18, 113-120. [https://doi.org/10.1016/S0928-0987(02)00251-8]

6. Peltonen, L. QbD approach for wet media milling of nanocrystals. Pharmaceutics 2018, 10, 104. [https://doi.org/10.3390/pharmaceutics10030104]

7. Li, M.; Yaragudi, N.; Afolabi, A.; Dave, R.; Bilgili, E. Sub-100 nm drug particles by wet milling with low bead contamination. Chem. Eng. Sci. 2015, 130, 207-220. [https://doi.org/10.1016/j.ces.2015.03.009]

8. Toneva, P.; Peukert, W. Modeling of mills and milling circuits. Handbook of Powder Technology 2007, 12, 873-911. [https://doi.org/10.1016/S0167-3785(07)12020-3]

9. Bilgili, E.; Guner, G. Mechanistic modeling of wet media milling for nanosuspensions. AAPS PharmSciTech 2021, 22, 1-23. [https://doi.org/10.1208/s12249-021-01950-7]

10. Randolph, A. Theory of Particulate Processes; Elsevier: 2012.

11. Ramkrishna, D. Population Balances; Academic Press: 2000.

12. Verkoeijen, D.; Pouw, G.A.; Meesters, G.M.; Scarlett, B. Population balances: A volume approach. Chem. Eng. Sci. 2002, 57, 2287-2303. [https://doi.org/10.1016/S0009-2509(02)00115-2]

13. Hounslow, M. The population balance for particle processes. KONA Powder Part. J. 1998, 16, 179-193. [https://doi.org/10.14356/kona.1998018]

14. Bilgili, E.; Hamey, R.; Scarlett, B. Nano-milling of pigments using a wet stirred media mill. Chem. Eng. Sci. 2006, 61, 149-157. [https://doi.org/10.1016/j.ces.2004.11.069]

15. Varinot, C.; Berthiaux, H.; Dodds, J. Prediction of product size in stirred bead mills. Powder Technol. 1999, 105, 228-236. [https://doi.org/10.1016/S0032-5910(99)00142-6]

16. Annapragada, A.; Adjei, A. Numerical simulation as an aid to process design of milling. Int. J. Pharm. 1996, 136, 1-11. [https://doi.org/10.1016/0378-5173(95)04484-2]

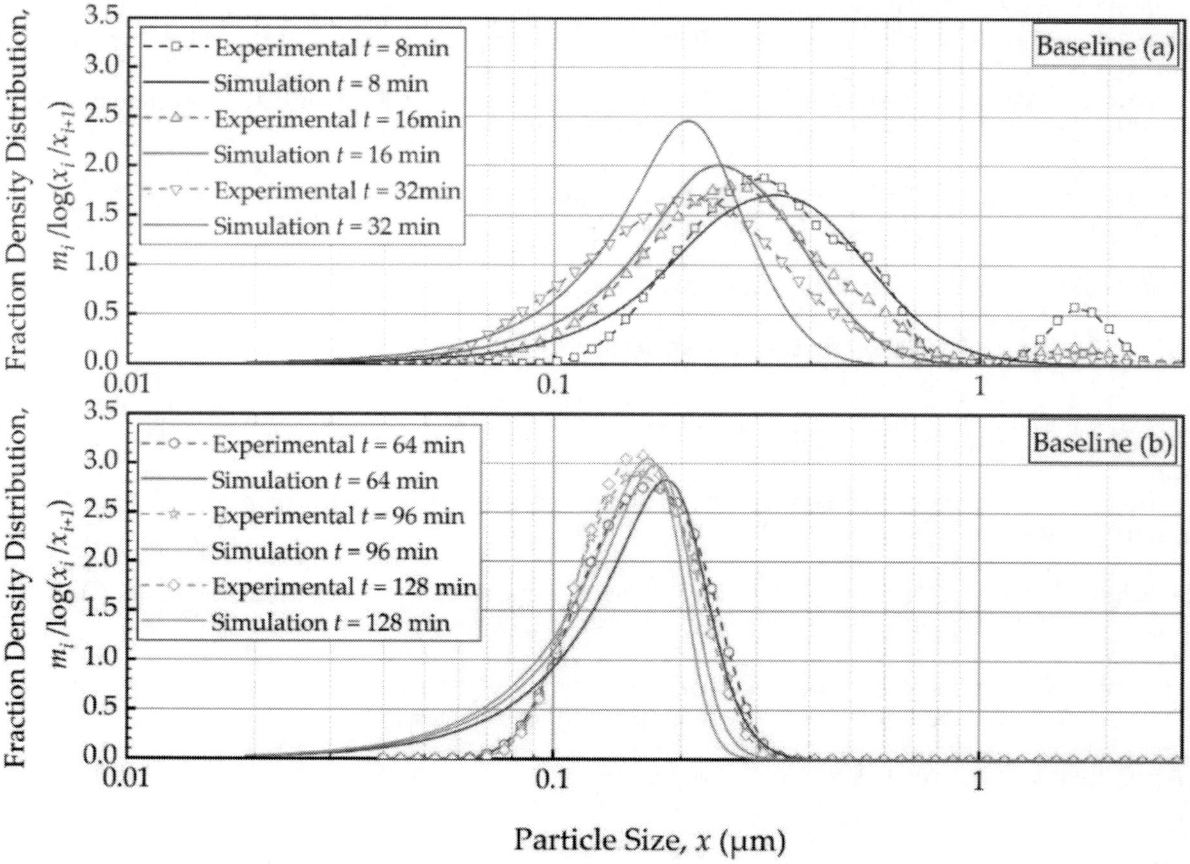

Figure 1 - Temporal variation of the mass fraction density distribution during (a) the initial 32 minutes of milling and (b) thereafter for the baseline experiment, with a volumetric flow rate of 126 mL/min and batch volume of 236 mL.

735ak Renewable Routes to Paracetamol: A Green Chemistry Analysis

Jimin Park, Caria Evans, Jacob Maier, Marta Hatzell, Carsten Sievers, Andreas Bommarius

Paracetamol is one of the most widely used active pharmaceutical ingredients (APIs) worldwide. Current methods of paracetamol production suffer from the use of petrochemical feedstocks, hazardous reagents, poor selectivity, and batch operation, leading to subpar Green Chemistry metrics. To build new processes to paracetamol, they must be renewable, safe, environmentally friendly, and continuous. Lignin derived phenol is a potentially renewable source of the raw material needed for paracetamol production.

A key difficulty in paracetamol synthesis is accessing the important intermediate 4-aminophenol, which is difficult to obtain selectively due to its instability. We have examined the viability of three routes to 4-aminophanol and paracetamol starting from phenol, a monomer product of lignin depolymerization, as the feedstock. These three routes are the acetamidation of hydroquinone, the imination and reduction of p-benzoquinone, and the hydrogenation of 4-nitrophenol. We examine the performance of each of these routes in regard to Green Chemistry metrics as well as viability for continuous synthesis. In addition, we compare the renewable routes with existing routes from petrochemicals.

Through our analysis, we identified that the 4-nitrophenol route is the most promising in terms of Green Chemistry metrics and continuous viability. Thus, we also examine the Green Chemistry metrics of different reactor configurations for the reaction, including a traditional batch reactor, a novel mechanochemical reactor, and a flow reactor.

735al Dissolution and Friability Analysis of Commercial and Lab-Developed Apap Tablets

Uche Chukwuemeka, Sheena Reeves, Asia Jackson

Previous research has sought to enhance fundamental knowledge in the area of particle interactions in acetaminophen (paracetamol, APAP) tablets by evaluating the performance of tablets formed using a single binding material or excipient. In comparison, commercially available brands of APAP may contain between 5-15 inactive ingredients. The selection of excipients can either improve or decrease tablet properties such as the bioavailability of the active pharmaceutical ingredients (API). To enhance patient safety, it is crucial to comprehend the relationship between APAP and various excipients. This research provides an analysis of commercial APAP tablets and lab-formed APAP tablets. The main goal is to create a drug that is easy for users to take without causing any adverse reactions to the non-medicinal ingredients. Dissolution and friability tests were used to measure tablet performance of regular strength (325 mg) including circle and oblong shapes. Tablet size and shape analysis was determined using Image Pro Premiere software. Most lab-formed tablets performed outside of the USP threshold of 1% max friability while the commercial brands had less than 0.5% friability. Disintegration experiments show an immediate loss of structure within the first 3 minutes for most tablets. The tablet shape, the amount of inactive ingredients, and the type of excipient affected the dissolution rate of the APAP tablet. The results emphasize the importance of continuous research to improve the composition and administration of APAP while addressing the changing sensitivities of patients worldwide.

735am Evaluation of Powder Triboelectric Properties: A Must Have for Material Optimization

Aurelien Neveu, Filip Francqui, Geoffroy Lumay, Marco Lupo

The charge buildup appearing during the processability of powders is a major problem in lot of industries involving the handling of powders. Indeed, during the flow of particulate materials the friction between the particles leads to the generation of electrostatic charges, a mechanism called tribocharging. As a consequence, a continuous accumulation of charges can appear inside the material, leading to numerous processability problems. Indeed, the charged particles have a high tendency to stick on the machine parts causing clogging of the pipes and time-consuming cleaning of the process line. Fine charged particles will also tend to agglomerate or stick on large grains, thus inducing a shift from the optimal particle size distribution defined at the powder formulation stage. The charge buildup will also result in the increase of the powder cohesiveness, electrostatic charges being one of the main cohesive forces, which is associated with a decrease of flowability. In some applications/materials, the presence of charges can also trigger dust explosions and thus must be avoided. Finally, with the development of continuous manufacturing, especially in the pharmaceutical industry, problems with charge buildup which where manageable for batch-to-batch processes become a major challenge as continuous manufacturing promotes the charge accumulation. Being able to accurately evaluate the sensitivity of a powder to tribocharging is thus a must have to move forward in the development of optimized processes and materials.

The tribocharging mechanisms are still not well understood, mainly due to their complexity. The charging ability of a powdered material is highly dependent on numerous properties such as the size and shape of the particles, the surface chemistry or the presence of water due to environment relative humidity. In addition, during the processing the powder interact with the machine parts (feeders, blades, pipes, ...), and the material of these pieces will also drastically change the tribocharging sensitivity. Therefore, the inherent complexity of tribocharging of powders prevent from using theoretical models for accurate prediction of the charge buildup.

But how can a powder formulator or process line designer evaluate and predict if the newly developed material will behave correctly during the production? One option is to produce a large batch and test it directly into the production line. However, this is a cost and time consuming process, especially with costly products like APIs. Fortunately, state of the art characterization methods have been developed during the last decade to assess this problematic. Especially, a powder tribocharging analyser (GranuCharge, Granutools, Belgium) allows to precisely evaluate the charge buildup during the flow. The instrument is composed of an electrometer allowing to accurately measure the net charge of the powder before and after the flow through a set of pipes. The material of the pipes can be changes to assess the influence of the process parts material on the charge buildup. The new characterization method opens up new possibilities for material optimization. In this presentation, we will show how the information gathered from this characterization method can be of great interest for tribocharging investigation. Several case studies for real applications will be presented to demonstrate the usefulness of the method.

735ap Smart Engineering Processes for Cbrn Protection: Configuration Pathways, Effectiveness and Limits

Youcef Serhane

Nuclear, radiological, biological and chemical threats still represent a major challenge for states, due to their potentially deleterious effects on the environment and human health. As these threats are unconventional, the solutions to deal with them must be unconventional in nature. In other words, it is not possible for a single discipline to find a solution to counter CBRN threat agents [1]. For this reason, and in order to deal with it, various emergency actions and procedures have been defined, notably through the implementation of major technological devices in the field of protection and decontamination techniques. These systems are based on effective filtration processes or radical destruction of toxic agents.

The aim of this investigation is to study the potential of combining certain processes for the elimination of toxic compounds, namely the transition from passive protection based on the principle of adsorption as a non-degradative capture process to active protection by combining the adsorbent filter with an advanced oxidation process [2][3]. A synergetic combination of these two processes with major constraints linked to the compactness and elimination performance of CBRN toxic compounds. Added to this are the problems of material recyclability and poisoning.

Furthermore, the study will also provide an opportunity to reflect on a solution adopted for the elimination of toxic agents, in particular through process intensification [4]. A solution that provides additional safety and greater effectiveness of the protection or decontamination system with regard to harmful compounds. In this study, the coupling of advanced oxidation processes, i.e. cold plasma and photocatalysis, will be discussed, with a focus on the reaction mechanism of chemical weapons (mustard gas, organophosphates and biological agents), the various limiting stages and also, to explore possible avenues of improvement, within the framework of practical implementation of the systems.

Reference:

[1] N.H. Johnson et al. INC, 2020. https://doi.org/10.1016/B978-0-12-819090-6.00002-7

[2] T. Rasheed et al. Chem. Environ. Eng. 2 (2020) 100037. https://doi.org/10.1016/j.cscee.2020.100037

[3] Y. Serhane et al. Chem. Eng. J. 471 (2023) 144326. doi.org/10.1016/j.cej.2023.144326

[4] W.A. Saoud et al. Applied Catalysis B: Environmental 241 (2019) 227–235. https://doi.org/10.1016/j.apcatb.2018.09.029

735aq Upscaling of Lithium-Ion Battery Models: From the Pore-Scale to the Cell-Scale through Homogenization.

Alessio Lombardo Pontillo, Agnese Marcato, Antonio Buffo, Gianluca Boccardo, Daniele Marchisio, Ilenia Battiato

Lithium-ion batteries (LiB) are currently the dominant energy storage system at many scales: from portable electronic devices to urban and aerospace vehicles, and likely they will play a major role in the construction of net zero energy buildings. Some studies have predicted a marked increment in the global battery demand of approximately 27% per year until 2030. Compared to alternative battery technologies, Li-ion batteries provide one of the best energy-to-weight ratios and experience low self-discharge when not in use. These properties, together with the decreasing production costs, have established Li-ion batteries as a leading candidate for the next generation of non-fossil fuel-powered systems [1].

There is a continued interest in the development of mathematical models that can efficiently and accurately describe the behavior observed during lithium-ion battery operation. Such models are often based on a simplified description of the electrochemical behavior because the morphological characterization of the electrode is complex and expensive from a computational point of view. However, batteries often exhibit non-uniform behavior in the current distribution, which can adversely affect battery performance, utilization, and lifetime. One of the aspects that is more responsible for this non-uniformity is the electrode morphology. Indeed, one of the limitations of the most common models that are present in literature is that each particle is reconstructed with a spherical shape. And if this approximation should be acceptable for the cathode side, the same can not be said for the anode side which is composed of graphite. This carbon form, which is the most stable one, is an anisotropic material because atoms are arranged in a hexagonal pattern forming a layer structure that makes graphite a good electric and heat conductor. Due to the layer structure, graphite particles have a flake shape, with one characteristic dimension smaller than the other two. The geometrical shape of graphite has a crucial impact on the whole electrochemical process involved in a lithium-ion battery and the correct reconstruction *in-silico* of the microscopic electrode geometry is fundamental to correctly understand the charge and discharge process from a macroscopical point of view [2].

Under these conditions, we decided to focus our attention on the geometrical characterization of the electrode of a graphitic half-cell. The main idea of this work is to start from a microscopic detailed model of a half-cell to obtain an upscaled model via homogenization techniques to speed up the simulation. This fast prediction is crucial to study in a short time different types of electrodes with a low computational cost.

To start, it was necessary to choose a way to reproduce graphite particles. To do this, we started looking for a geometric shape that met several requirements. The shape should be a compromise between accuracy and computational cost. An extremely realistic representation, leading to an equally highly complex electrode geometry, would be very difficult to recreate, owing to the necessity of using computationally expensive geometric algorithms, and it would also lead to expensive battery simulations. On the other hand, using too simplified of a model (e. g. spheres as previously mentioned) would lead to an unacceptably unrealistic electrode representations, impacting the quality of the simulation results. After these considerations, we decided to adopt an ellipsoidal shape for the particles. Each particle is characterized by a semi-axis much shorter than the other two. The two larger characteristic lengths are taken from particle size distribution (PSD) data obtained in laboratory tests, together with an evaluation of the particles aspect ratio from SEM image analysis. In this way, we were able to obtain a good representation of the electrode geometry, and this was confirmed by the experimental evaluation of the porosity. The packing of ellipsoids was produced with a combination of two Python based open-source codes. The first one is Yade-DEM and we used it to generate a package of spheres starting from the experimental PSD. Later, each sphere was transformed into a flattened ellipsoid with the software Blender. Then, a rigid-body simulation was conducted to compact the layer of particles and to remove the empty space formed after the transformation of each sphere. Finally, the central portion of the package was extracted and used as input to setup the pore-scale simulations, performed with the software COMSOL 6.1. This whole process is governed by a Python script that allows us to produce different electrodes in a semi-automatic way.

The same script was also used to create the model used for the homogenization process. In this case there are some important conditions that should be respected. The main idea of homogenization is to find the macroscopic properties by investigating microscopical behavior. This upscaling is possible thanks to the multiple-scale expansion. This technique consists in rewriting the equations in a dimensionless form to extract, in this case, the Damköhler number and the Péclet number. These two parameters define the condition in which there is scale separation and the possibility of homogenizing microscopic equations. Then, the solution of a closure problem on a periodic cell is necessary to obtain the upscaled effective values for both the diffusivity coefficient and the electrical conductivity for the liquid and solid phases [3-4]. The closure problem is solved on a unitary periodic cell that has a dimension many times smaller than the pore-scale model, providing for a microscopic representative volume. For the case under consideration, the unitary cell has been created with a central core of polydisperse ellipsoids, and a few semi-ellipsoids arbitrarily placed on the boundaries to respect the unit cell periodicity. Once obtained the effective values for the diffusion coefficient and for the electrical conductivity, is possible to create and solve the homogenized model.

The results extracted from the homogenized model are then compared and validated with the ones obtained from the 4D pore scale simulation. With the homogenized model we can obtain the same results as the pore scale model but in shorter times and less computational resources. With the script created for the electrode geometry, we can obtain different morphologies and particles size distribution. The flexibility of the model, coupled with the possibility of studying the behavior of different electrodes, will allow for the easy creation of a large dataset on a wide parameter space, to be used as training data for a neural network model, used a quick surrogate model to expand the range of explorable electrode configurations.

[1] McKinsey & Company, **(2023)**, *Battery 2030: Resilient, sustainable, and circular,* https://www.mckinsey.com/industries/automotive-and-assembly/our-insights/battery-2030-resilient-sustainable-and-circular

[2] Lu, X., Lagnoni, M., Bertei, A. *et al.,* **(2023)**, *Multiscale dynamics of charging and plating in graphite electrodes coupling operando microscopy and phase-field modelling.* Nat Commun 14, 5127.

[3] Arunachalam, H., Onori, S., Battiato, I., **(2015)**, *On Veracity of Macroscopic Lithium-Ion Battery Models. J. Electrochem. Soc.,* 162, A1940

[4] Ling, B., Battiato, I., **(2020)**, *τ-SIMPLE Algorithm for the closure problem in homogenization of stokes flows.* Advances in Water Resources, 144, 103712.

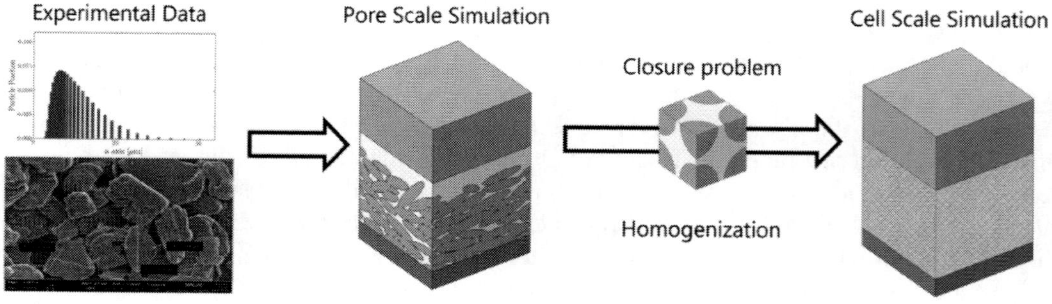

735ar A Systematic Model-Based Estimation of State of Health and State of Charge for Second-Life Li-Ion Batteries

Md Emdadul Haque, Auggie Chico, Greg Lusk, John Kolar, Paramarshi Banerjee, Anurag Srivastava, Debangsu Bhattacharyya

Electric vehicles (EVs) are now being widely used as environmentally friendly alternatives to traditional combustion-engine vehicles. Lithium-ion (Li-ion) batteries are widely adopted as the main power source for EV's. Their popularity stems from their high energy density, environmental friendliness, and long-life cycle, making them the preferred choice in the EV industry [1]. However, the energy storage capacity of these batteries decreases over the battery lifetime because of aging. Therefore, the state of battery aging is assessed for safe and smooth driving experience [2]. Once an appropriate aging indicator reaches the predetermined limit, the battery is to be replaced [3]. As a result, there is a surge in battery or electro-chemical waste, necessitating the development of technologies to handle these batteries once they reach the end of their original application. Creating new technologies for the safe and useful repurposing of batteries is crucial to mitigate severe environmental and economic consequences and for maximum utilization of these batteries. This study considers second-life use of these batteries for stationary applications such as for storing renewable energy. Even for those applications, a used battery's state of health (SOH) and state of charge (SOC) should be accurately estimated for reliability of the energy storage systems. Furthermore, it is desired that their remaining useful life (RUL) be predicted so that replacement/maintenance planning can be undertaken without compromising system integrity and reliability. Existing approaches to the monitoring of SOH and SOC and prediction of RUL of Li-ion batteries can be broadly classified into three main approaches: (1) heuristics-based methods, (2) model-based techniques, and (3) data-driven approaches [4,5]. This work proposes a hybrid approach where a detailed first-principles electrochemical model is developed and integrated with real-time data to optimally adapt model parameters and update state estimates. The current estimates of state and model parameters are then utilized to estimate the current SOH and SOC and predict RUL by projecting current states and parameters into the future. In this hybrid approach, the electrochemical model computes the spatial distribution of electrolyte concentration based on the concentrated solution theory, along with the potential distribution in the solid and liquid phases across the anode, cathode, and separator. Several approaches are developed in this study, including model reformulation, coordinate transformation, orthogonal collocation, and recursive Bayesian estimation for computational tractability of the optimal estimation problem for this spatially distributed large-scale partial differential algebraic equation system. The approach is evaluated by applying to the retired batteries collected from EVs.

References:

1. Shah, F.A.; Sheikh, S.S.; Mir, U.I.; Athar, S.O. Battery Health Monitoring for Commercialized Electric Vehicle Batteries: Lithium-Ion. *Proc of the 5th International Conference on Power Generation Systems and Renewable Energy Technologies (PGSRET), Istanbul Technical University, Turkey*, **2019**.
2. Xiong, R.; Li, L.; Tian, J. Towards a smarter battery management system: A critical review on battery state of health monitoring methods. *Power Sources*, **2018**, 405, 18-29.
3. Berecibar, M.; Gandiaga, I.; Villareal, I.; Omar, N.; Van Mierlo, J.; Van Den Bossche, P. Critical review of state of health estimation methods of Li-ion batteries for real applications. *Sustain. Energy Rev.*, **2016**, 56, 572-587.
4. Wang, F.; Zhao, Z.; Ren, J.; Zhai, Z.; Wang, S.; Chen, X. A transferable lithium-ion battery remaining useful life prediction method from cycle-consistency of degradation trend. *Power*

Sources, **2022**, 521, 230975.

5. Jin, S.; Sui, X.; Huang, X.; Wang, S.; Teodorescu, R.; Stroe, D. Overview of Machine Learning Methods for Lithium-Ion Battery Remaining Useful Lifetime Prediction. *Electronics*, **2021**, 10(24), 3126.

735as Advancing High Energy Dense Li-S Pouch Cell through Form Factor Optimization

Sayan Das, MSA Bhuyan, Krish Naresh Gupta, Omena Okpowe, Austin Choi, David Olawale, Vilas Pol

The success of lithium-ion batteries (LIBs) has been the catalyst for a revolution in electrochemical energy storage and the main driving force for widespread adoption of mobile devices. Sulfur is abundant, low-cost, and environmentally friendly, making it a promising alternative to LIBs. Battery technology with high energy density is being driven by the shift to renewable energy sources and the electrification of the transportation sector. The high energy density of lithium-sulfur (Li-S) batteries[1] (2,600 Wh kg^{-1}) makes them a viable prospect for the next generation of energy storage. However, the practical energy yield is lower than theoretically predicted because of problems like lithium polysulfide (LiPS) shuttling and Li dendrite growth. Advanced mitigation strategies are needed because these problems result in rapid capacity fade, reduced coulombic efficiency, and potential safety hazards.

Recent advancements in Li-S technology have led to an increase in sulfur content within electrodes, highlighting the positive impact of carbon materials on enhancing electrochemical performance. However, challenges persist, including the considerable 'dead weight' from inactive components[2] that significantly detracts from the batteries' specific energy. Efforts to reduce the electrolyte volume are constrained by the reactive nature of lithium metal, the fragility of the solid electrolyte interface (SEI), and the necessity for conductive carbon and sufficient electrolyte to support cathode functionality.[3] The design parameters for Li-S batteries, especially in pouch formats, markedly deviate from those of coin cells, influencing their cycle life and overall stability. The endeavor to diminish the electrolyte volume relative to the sulfur load (E/S) and the negative to positive capacity (N/P) ratios[4] for pouch cells introduces fresh challenges and highlights the complexity of producing stable, long-lasting Li-S batteries on a large scale.

Pouch cells of Li-S batteries with capacities ranging from 375 mAh to 1.4 Ah have been developed[4]. Our findings indicate that sulfur loading, and the electrolyte/sulfur ratio are essential form factors for achieving high energy density batteries in a pouch cell format. We have engineered a 1.45 Ah pouch cell with optimized sulfur loading and an electrolyte/sulfur ratio, attaining an energy density of 151 Wh/kg and a capacity retention rate of 70% for up to 100 cycles. Additionally, the developed 1Ah pouch cell exhibits a 68% capacity retention after 50 cycles, aiming for a higher energy density of 180 Wh/kg. Morphological analysis of cycled electrodes reveals that the primary cause of pouch cell failure[4] is attributed to Li metal powdering and subsequent polarization, rather than LiPS shuttling. To prove real-world viability, Li-S pouch cells have been successfully integrated into a functional drone.[4]

References

1. W. Weng, V. G. Pol, K. Amine, 'Ultrasound Assisted Design of Sulfur/Carbon Cathodes with Partially Fluorinated Ether Electrolytes for Highly Efficient Li/S Batteries,' *Adv. Mater.*, 2013, 25, 1608–1615.

2. C. N. Hong, D. K. Kye, A. U. Mane, V. Etacheri, V. G. Pol, "Blocking Polysulfides in Graphene–

Sulfur Cathodes of Lithium–Sulfur Batteries through Atomic Layer Deposition of Alumina", *Energy Technology*, 2019, 1900621

3. A. D. Dysart, N. A. Cardoza, G. Mitchell, V. Ortalan, V.G. Pol, "Effect of Synthesis Method Using Varying Types of Micropore Level Sulfur Infiltration on Electrochemical Performance in Lithium–Sulfur Batteries," *Energy Technology*, 2019, 1900194.

4. S. Das, MSA Bhuyan, K. N. Gupta, O. Okpowe, A. Choi, J. Sweeny, D. Olawale, V. G. Pol, Optimization of the Form Factors of Advanced Li-S Pouch Cells, *Small* 2024, 2311850

735at Gas Diffusion Electrode Improves Kinetics of Rechargeable Aluminum-CO_2 Batteries.

Gustavo Diaz, Shuya Wei

Aluminum (Al) is a very attractive anode material for electrochemical CO_2 conversion due to its lower cost and reactivity in comparison to other battery anode materials including lithium, sodium, potassium, etc. Additionally, the high specific energy of aluminum (2980A·hour/kg) suggests that large amounts of electrical energy are possible during the electrochemical conversion with CO_2. Primary Al-CO_2 batteries have a remarkable discharge capability with the introduction of a small amount of oxygen. Unfortunately, these batteries are not yet rechargeable. By utilizing a homogeneous iodine-based redox mediator, our group demonstrated the reversibility of the Al-CO_2 battery. This secondary battery can achieve discharge and charge with a minimal overpotential of 0.05V. The present-day rechargeable Al-CO_2 battery maintains high discharge capacity and enables the battery to be recharged for up to 12 cycles at a 20 mA/gcarbon rate.

To achieve a higher cyclability on Al-CO_2 batteries, modifications regarding the hydrophobicity, porosity, and high surface area of the gas diffusion electrode (GDE) were explored. By utilizing a mixture of two-dimensional and three-dimensional high surface area carbon-based cathode materials (Graphene and Ketjenblack EC-600JD) with unique pore structures, we demonstrate that the rechargeable Al-CO_2 battery can enhance CO_2 reduction during the discharge processes.

In addition, numerous electrochemical evaluations including cycle voltammetry (CV), impedance (PEIS), and galvanostatic charge and discharge (GCD) were conducted to examine the dependency between the active species in the electrolyte and the active cathode material. During the GCD test, an enhancement of 154 cycles at 100 mA/g was observed when doubling the electrolyte volume and incorporating an external CO_2 tank into the system. The CV examinations showed an increase in the charge and discharge from 0.4mA to 0.92mA, and 0.2mA to 0.45mA respectively. The integration of the additional electrolyte, the CO_2 external tank, and the new GDE surpasses the battery cyclability over 1250 percent in comparison to the original setup. These examinations utilizing this new cathode prove that the Al-CO_2 battery exhibits a notable cycle stability and current density improvement.

STUDY OF GALVANIC CORROSION OF MILD STEEL COATED WITH AZ91D MAGNESIUM ALLOY BY COMSOL MULTIPHYSICS

Waliul Islam Khan [1], Sk. Moumita Manjur [3], Gautam Gupta [2], Sidney Lin [3]

[1]Department of Chemistry, University of Louisville, Louisville. KY, USA

[2]Department of Chemical Engineering, University of Louisville, Louisville, KY, USA

[3]Department of Chemical Engineering, Lamar University, Beaumont, Texas, USA

Abstract

Pipeline corrosion is associated with a global economic loss of $1.3 billion yearly, also it increases risk of spillage of flammable and toxic substances that can result in significant safety hazard. Pipeline corrosion was studied by analyzing galvanic corrosion between pipe steel and AZ91D magnesium alloy coating. It was found that, as the radius of the hollow and the total radius of the pipe decreased and as the magnesium alloy coating thickness increased, corrosion rate and current density decreased, and galvanic corrosion potential increased at the outer surface of the magnesium coating however the changes are not pronounced in intersection between steel and magnesium coating.

Introduction

Corrosion is the process of gradual deterioration of the materials by electrochemical or environmental factors. Corrosion is a challenging phenomenon in terms of performance and longevity of advanced technologies including electronics, medical implants and renewable energy systems. One of the most common types of corrosion is galvanic corrosion. Galvanic corrosion happens when two metals encounter each other in electrolyte, one metal can be considered as anode and the other metal can be considered as a cathode based on the potential of these two metals. In this case, electron moves from anode to cathode and anode gets corroded by being dissolved in the electrolyte. The rate of the galvanic corrosion can be calculated based on the equation-

$$I_g = (E_c - E_a)/(R_a + R_c + R_m + R_s) \ [1]$$

Here, galvanic corrosion is determined by I_g, E_c is the equilibrium potential of cathode and E_a is the equilibrium potential of anode, R_a is the resistance for anode, Rc is the resistance for cathode, R_s is the solution resistance, R_m is the resistance from anode to cathode.[1]

In the galvanic corrosion process the metal which is less noble, works as a sacrificial anode, which protect the more noble metal that work as a cathode. Only when the anode gets completely dissolved into the electrolyte, cathode tends to rust. For example, Iron can rust only when zinc coating is totally corroded, in the event of iron coated with zinc immersed in an electrolyte solution. Galvanic corrosion has been studied by many researchers based on different parameters. Galvanic corrosion between galvanized steel and mild steel at different pH was studied by Vagge *et.al.*,[2] change of corrosion rate with electrolyte thickness between steel bolt and plate joint had been studied by Essali.M *et.al.*,[3] the presence of two microstructures in magnesium alloy was studied by Deshpande.KB,[4] galvanic corrosion between magnesium and aluminum was studied by Loic Lacroix *et.al.*,[5] Nicolas Murer *et.al.*[6] observed that due to the presence of insulator between aluminum alloy galvanic corrosion rate increased. Similar insulator edge effect was also observed in the galvanic corrosion study between magnesium alloy (AE44) and mild steel done by Dao Trinh *et.al.* and it was found that due to the presence of insulator galvanic corrosion rate increased.[7] In the galvanic corrosion study between rivets made of steel joint and aluminum plate, Y.Zhang *et.al.* observed that corrosion rate was maximum around the joint and corrosion rate was lowest at the edges.[8] One of the most convenient way to measure galvanic corrosion rate is using simulation procedure. In this research work galvanic corrosion between pipe steel and AZ91D magnesium alloy coating has been studied by a finite element analysis (FEA) mathematical model developed using COMSOL Multiphysics 4.2.

Materials and methods

Galvanic corrosion between pipe steel and AZ91D magnesium alloy coating immersed in 5 wt% aqueous NaCl solution which had a conductivity of 7.9500 S/m and height of the solution was 7.5 cm was studied by COMSOL Multiphysics. 10-inch pipeline of schedule 40 and 80 and 12- inch pipeline of schedule 40 and 80 and magnesium coating of 8 different thickness were used for the study.

Table 1. AZ91D Mg alloy has the percentage composition as following [9]

Aluminium	8.3-9.7
Copper	0.03 max
Magnesium	88.6-91.04
Iron (max)	0.005
Nickel (max)	0.002
Zinc	0.35-1.0
Manganese	0.15-0.5
Silicon	0.1 max
Other metallic	0.02

Table 2. The percentage chemical composition of mild steel is as following [10]

Carbon	0.16-0.18
Silicon	0.40 (max)
Manganese	0.70-0.90
Sulphur	0.040 (max)
Phosphorous	0.040 (max)
Iron	98.44-98.66s

For different values of radius of hollow radius, radius of outer radius of the pipeline and different values of the magnesium coating thickness simulated values of the variation in electrode potential and current density were obtained for the galvanic couple. Maximum values for the current densities and potential were considered for analyzing the corrosion behavior of the system.

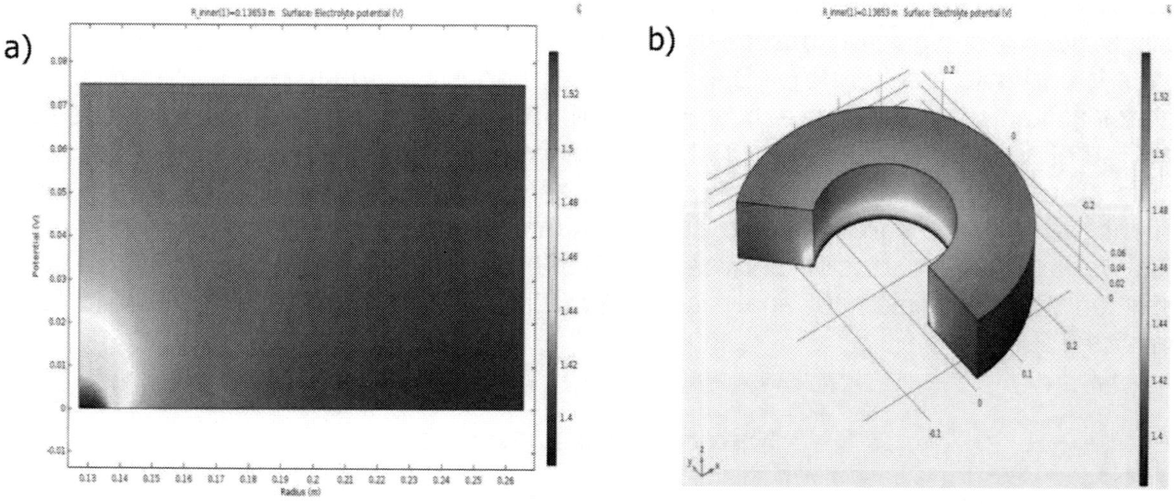

Figure 1. Variation in the electrode potential with radius in the galvanic couple of stainless steel and magnesium coating a) 2D image, b) 3D image

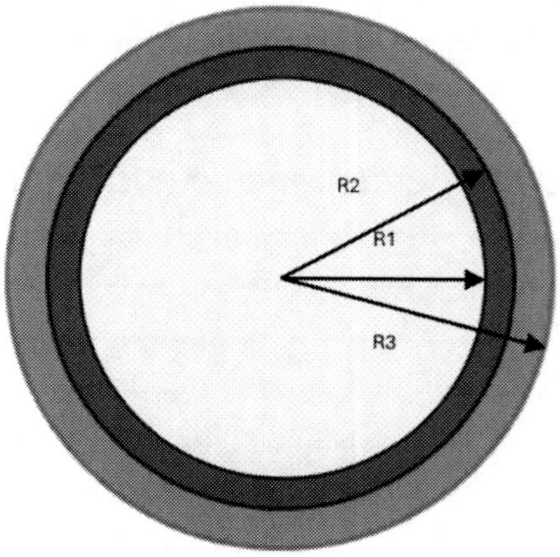

Figure 2. Schematic diagram of steel pipeline with magnesium coating

The schematic diagram in Figure 2 is showing the pipeline image where R1 is representing the radius of the hollow, R2 is representing the outer radius and R3 is representing the total radius, that is the outer radius plus the radius of the magnesium coating.

Table 3. The following 32 experiments were performed for the pipelines of 10 inch and 12 inch having 40 and 80 schedule number respectively and for eight different magnesium coating thickness.

Test no	Nominal pipe size (inch)	Outer Diameter (inch)	Inner Diameter (inch)	Schedule number	Thickness of Mg coating (inch)	R1 (m)	R2 (m)	R3 (m)
1	10	10.75	10.02	40	0.0625	0.127254	0.136525	0.265367
2	10	10.75	10.02	40	0.125	0.127254	0.136525	0.266954
3	10	10.75	10.02	40	0.1875	0.127254	0.136525	0.268542
4	10	10.75	10.02	40	0.25	0.127254	0.136525	0.270129
5	10	10.75	10.02	40	0.3125	0.127254	0.136525	0.271717
6	10	10.75	10.02	40	0.375	0.127254	0.136525	0.273304
7	10	10.75	10.02	40	0.4375	0.127254	0.136525	0.274892
8	10	10.75	10.02	40	0.5	0.127254	0.136525	0.276479
9	10	10.75	9.562	80	0.0625	0.121437	0.136525	0.25955
10	10	10.75	9.562	80	0.125	0.121437	0.136525	0.261137
11	10	10.75	9.562	80	0.1875	0.121437	0.136525	0.262725
12	10	10.75	9.562	80	0.25	0.121437	0.136525	0.264312
13	10	10.75	9.562	80	0.3125	0.121437	0.136525	0.2659
14	10	10.75	9.562	80	0.375	0.121437	0.136525	0.267487
15	10	10.75	9.562	80	0.4375	0.121437	0.136525	0.269075
16	10	10.75	9.562	80	0.5	0.121437	0.136525	0.270662
17	12	12.75	11.938	40	0.0625	0.151613	0.161925	0.315125
18	12	12.75	11.938	40	0.125	0.151613	0.161925	0.316713
19	12	12.75	11.938	40	0.1875	0.151613	0.161925	0.3183
20	12	12.75	11.938	40	0.25	0.151613	0.161925	0.319888
21	12	12.75	11.938	40	0.3125	0.151613	0.161925	0.321475
22	12	12.75	11.938	40	0.375	0.151613	0.161925	0.323063
23	12	12.75	11.938	40	0.4375	0.151613	0.161925	0.32465
24	12	12.75	11.938	40	0.5	0.151613	0.161925	0.326238
25	12	12.75	11.374	80	0.0625	0.14445	0.161925	0.307962
26	12	12.75	11.374	80	0.125	0.14445	0.161925	0.30955
27	12	12.75	11.374	80	0.1875	0.14445	0.161925	0.311137
28	12	12.75	11.374	80	0.25	0.14445	0.161925	0.312725
29	12	12.75	11.374	80	0.3125	0.14445	0.161925	0.314312
30	12	12.75	11.374	80	0.375	0.14445	0.161925	0.3159
31	12	12.75	11.374	80	0.4375	0.14445	0.161925	0.317487
32	12	12.75	11.374	80	0.5	0.14445	0.161925	0.319075

Results and discussions

For 8 different magnesium coating thickness change of current density, electrochemical potential and corrosion rate were analyzed for pipeline of 10-inch length and pipeline of 12-inch length having schedule 40 and 80 respectively.

It was found that, as the total radius of the pipe and the magnesium alloy coating thickness increased and the hollow radius increased, current density decreased in the outer surface of the magnesium coating, however the changes are not pronounced at the at the intersection between steel and magnesium coating which can be seen in Figure 3

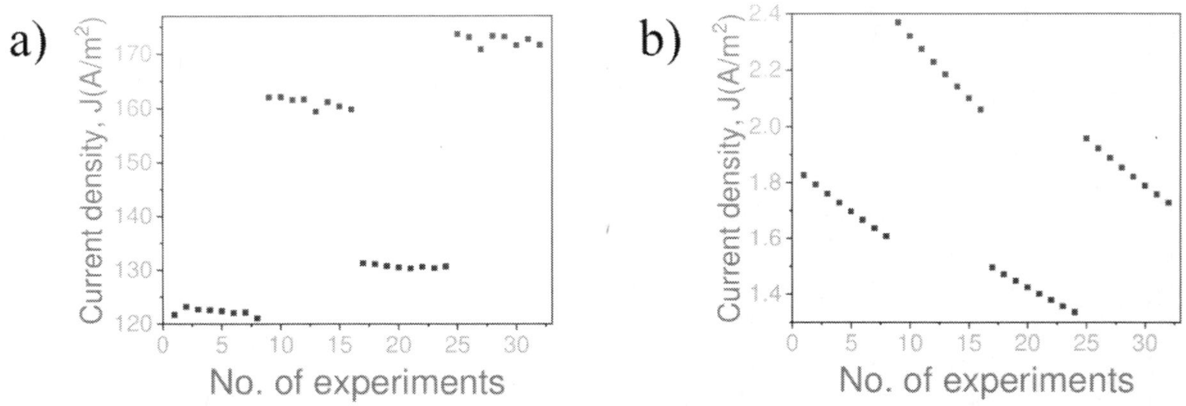

Figure 3. Current density a) at the intersection between steel and magnesium coating and current density b) at the outer surface of magnesium coating

Again since there is a linear relationsip between current density and corrosion rate [11], it was observed that, as the total radius of the pipe and the magnesium alloy coating thickness increased and the hollow radius increased, corrosion rate decreased in the outer surface of the magnesium coating, however the changes are not pronounced at the intersection between steel and magnesium coating that is shown in Figure 4.

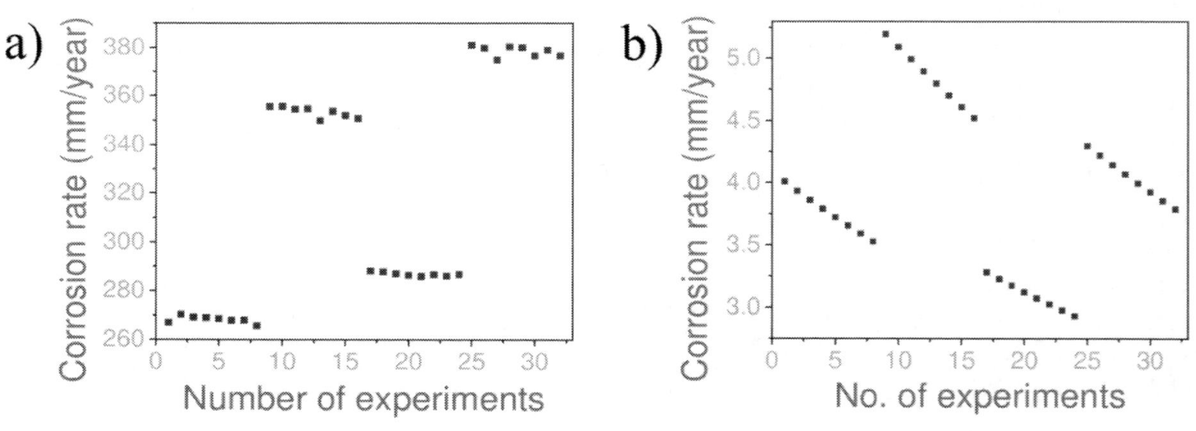

Figure 4. Corrosion rate a) at the intersection between steel and magnesium coating and corrosion rate b) at the outer surface of magnesium coating

Since, due to the polarization effect with the increase of galvanic potential current density and corrosion rate decreases it was observed that, as the total radius of the pipe and the magnesium alloy coating thickness increased and the hollow radius decreased, galvanic potential increased in the outer surface of the magnesium coating, however the changes are not pronounced at the at the intersection between steel and magnesium coating that can be seen in Figure 5.

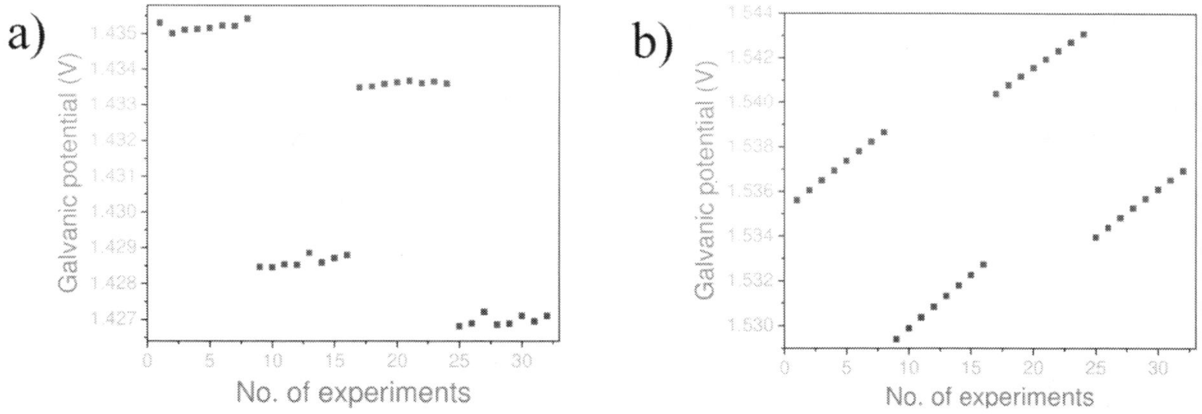

Figure 5. Galvanic potential a) at the intersection between steel and magnesium coating and b) galvanic potential at the outer surface of magnesium coating

Conclusion

Change of current density, corrosion rate and galvanic potential for the galvanic corrosion between pipe steel and AZ91D magnesium alloy coating have been analyzed in this research article by varying different hollow radius, outer radius and magnesium coating thickness for the pipeline of size 10 inch and 12 inch and schedule no. 40 and 80 respectively. It was found that with the increase of total pipeline radius and magnesium coating thickness and the hollow radius decreased, current density and corrosion rate increases and galvanic corrosion rate decreases in the outer surface of the magnesium coating. This information can be useful for predicting lifeline of pipeline which can be useful for safety and economic purpose.

Acknowledgement

The authors like to acknowledge for travel support by the project funded by U.S. Department of Energy, Office of Science, Biological and Environmental Research Program under award number DE-SC0023216.

References

1. Song, G.; Johannesson, B.; Hapugoda, S.; StJohn, D.(2004) Galvanic corrosion of magnesium alloy AZ91D in contact with an aluminium alloy, steel and zinc. Corrosion Science, 46 (4), 955-977

2. Kamble, P. A.; Deshpande, P.; Vagge, S.(2022) Numerical investigation of galvanic corrosion between galvanized steel and mild steel in bolted joint. Materials Today: Proceedings, 50, 1923-1929.

3. Radouani, R.; Echcharqy, Y.; Essahli, M.(2017) Numerical simulation of galvanic corrosion between carbon steel and low alloy steel in a bolted joint. International Journal of Corrosion, 2017 (1), 6174904.

4. Deshpande, K. B. Numerical modeling of micro-galvanic corrosion.(2011) Electrochimica Acta, 56 (4), 1737-1745.

5. Lacroix, L.; Blanc, C.; Pébère, N.; Tribollet, B.; Vivier, V.(2009) Localized approach to galvanic coupling in an aluminum–magnesium system. Journal of the Electrochemical Society, 156 (8), C259.

6. Murer, N.; Missert, N.; Buchheit, R.(2008) Towards the modeling of microgalvanic corrosion in aluminum alloys: the choice of boundary conditions. In Proceedings of the COMSOL Users Conference Boston.

7. Trinh, D.; Dauphin Ducharme, P.; Mengesha Tefashe, U.; Kish, J. R.; Mauzeroll, J.(2012) Influence of edge effects on local corrosion rate of magnesium alloy/mild steel galvanic couple. Analytical chemistry, 84 (22), 9899-9906.

8. Zhang, Y.; Chen, Y. L.; Fan, W. J.; Wang, C. G.; Bian, G. X.(2017) Study on the galvanic corrosion of aluminum alloy single bolted lap joints in simulated atmospheres. Materials and Corrosion, 68 (10), 1107-1115.

9. Dynacast. Magnesium casting metals. Dynacast, (2023). https://www.dynacast.com/en/knowledge-center/material-information/magnesium-casting-metals (accessed 10 November 2024).

10. LIMITED, K. S. S. *Mild steel.* KV STEEL SERVICES LIMITED, (2009-2024). https://kvsteel.co.uk/steel/mild-steel.html (accessed 10 November 2024).

11. Lu, B.; Luo, J.; Mohammadi, F.; Wang, K.; Wan, X.(2008) Correlation between repassivation kinetics and corrosion rate over a passive surface in flowing slurry. Electrochimica Acta , 53 (23), 7022-7031

735aw Unusual Transport of Phenolic Compounds in Polymer Membranes.

Woo Jin Jang

Bio-oil, the product of fast pyrolysis of biomass, can be further processed to a value-added energy source, such as biodiesel. Membrane-based separation has been proposed as a potential separation technology for this sustainable feed. As a model system, we explore a separation of a mixture of water, methanol, and guaiacol via a polymer membrane. We observe that the flux of the components in these systems are highly coupled, leading to unusual observations in the membrane separation performance.

Estimating Prediction Uncertainty of a Data-Driven Plant Model for Control Applications

Nicolas Blum, Valentin Krespach, Sebastian Rehfeldt, Harald Klein

The need for load flexibilization and increased efficiency in energy-intensive processes has become increasingly relevant in recent years, driven in part by the energy crisis. In addition to optimizing the plant design, the control of process variables plays a decisive role.

Linear model predictive controllers (LMPC) currently in widespread use can only partially map the often highly non-linear, thermodynamic processes that occur in air separation plants (ASU), for example. In the last ten years, the field of model-based reinforcement learning (RL) has evolved approaches that can learn improved non- linear control without prior physical knowledge, but purely through historical data and interaction with the dynamic process system. One main component of such RL-based controllers are data-driven process models. The reliability of the model prediction is crucial for the performance of the resulting controller.

Deep neural networks (DNN) are commonly used in the RL domain and have shown good data fitting and generalization capabilities. However, besides their advantages, data-driven models are also known to lack generalization performance within operation regions that were not covered by the training dataset, leading to potential instabilities of the resulting controller and the process behavior.

The aim of this work is to show one possible approach to estimate the uncertainty expressed as standard deviation of the model's expected prediction. This additional information can then be considered by the controller's planner to mitigate prediction uncertainty during optimizing control decisions. The uncertainty estimation of the used approach is investigated with respect to the prediction horizon and coverage of data of a specific operation region.

735az Selective Separation of Precious Ions from Water By Mxene Free-Standing Membranes

Ahmed Al Mayyahi, Milad Esfahani

Titanium carbide MXene free-standing membranes (MFMs) have garnered significant attention for their efficiency in ion separation from water, owing to their aligned nanochannels and unique surface chemistry. However, their application in the selective separation of precious ions is seldom and remains to be experimentally explored. In this study, we focused on adjusting the interlayer spacing of MFMs and, consequently, the size of their nanochannels to either match or exceed the diameter of the desired ions. This adjustment facilitates the permeation of precious ions, including Y^{2+}, Cr^{2+}, and Ce^{2+}, through the membrane while rejecting other ions, allowing for their selective separation. We investigated the correlation between membrane surface chemistry and ion selectivity, showcasing the potential of controlling MXene surface chemistry to enhance the efficiency of selective ion separation. Additionally, we extended the lifetime and stability of MFMs through in situ cross-linking via thermal treatment. Our findings pave the way for the practical implementation of these membranes in the selective separation of precious ions, a crucial process for technologies essential to fostering a circular economy and achieving a clean energy future.

735ba A Population Balance Based Model to Optimize Coagulation-Flocculation

Diogo Abreu, Pablo Gallo, David Fernandes del Pozo, Ingmar Nopens

Given the decreasing availability of drinking water, the growing needs of the global population, climate change, and desertification, drinking water treatment is increasingly necessary. The conventional drinking water treatment train comprises four steps: coagulation-flocculation, sedimentation, filtration, and disinfection. During the first step of the process, colloidal particles bind to each other and form flocs, whose physical properties, such as size and shape, determine their settleability and filterability. The success of subsequent steps will be mainly driven by these characteristics. Therefore, coagulation-flocculation can be deemed the essential component of the conventional drinking water treatment scheme, as it affects the next steps of the treatment and its efficiency strongly impacts the performance of the overall treatment (Tzoupanos et al., 2008).

In view of the above, the optimization of coagulation-flocculation is fundamental to optimizing drinking water treatment. Coagulant dosage is considered to be the most significant of the several variables that are known to influence this process, because it is largely related to ensuring an efficient and cost-effective treatment. Given its importance, the main focus of optimization strategies is thus the determination of the optimal coagulant dosage. Currently, jar tests, which are lab-scale simulations of the coagulation-flocculation and sedimentation stages, and/or machine learning models are used to identify the optimal coagulant dosage (Ratnaweera & Fettig, 2015). However, both methods have limitations that hamper their ability to successfully optimize coagulation-flocculation. The former is time-consuming and cannot cope with rapid changes in process conditions in real time, while the latter struggles with unforeseen events and can be difficult to interpret.

It is hypothesized that a potential way forward to improving the prevalent optimization techniques is the incorporation of floc properties (Khedher et al., 2023). To explicitly consider them in the optimization strategy, the Population Balance Modelling (PBM) framework is the ideal tool. This type of model describes the evolution of particulate systems suspended in a continuous phase by tracking the dynamics of key properties, such as size distribution and shape. The successful development of a PBM model could potentially lead to improvements in both state-of-the-art optimization strategies. This approach could be used to reduce the number of jar tests needed to identify the optimal coagulant dosage or even to replace them. Moreover, the PBM could be used to integrate mechanistic-based knowledge into the machine learning based approaches to improve interpretability and their capability of dealing with unforeseen events.

In this work, we present the first population balance based model capable of determining the optimal coagulant dosage. The model was calibrated and validated on an extensive data set obtained by the authors. The experiments performed describe the dynamic evolution of flocs' particle size distribution and mass fractal dimension throughout a jar test for multiple coagulant dosages. The goal of this modelling exercise is to show the importance of considering floc properties in the optimization process, as well as deepen the current understanding of how these properties evolve and impact the particle separation stages of the treatment. Finally, this work seeks to develop a novel optimization strategy that could potentially be used as a complement or an alternative to the prevailing optimization methods.

References

Khedher, M., Awad, J., Donner, E., Drigo, B., Fabris, R., Harris, M., Braun, K., & Chow, C. W. K. (2023). Using the Flocculation Index to optimise coagulant dosing during drinking water treatment.

Journal of Water Process Engineering, *51*. https://doi.org/10.1016/j.jwpe.2022.103394

Ratnaweera, H., & Fettig, J. (2015). State of the art of online monitoring and control of the coagulation process. In *Water (Switzerland)* (Vol. 7, Issue 11, pp. 6574–6597). MDPI AG. https://doi.org/10.3390/w7116574

Tzoupanos, N., Zouboulis, A., Tzoupanos, N. D., & Zouboulis, A. I. (2008). Coagulation-flocculation processes in water/wastewater treatment: the application of new generation of chemical reagents Coagulation-Flocculation, 6th IASME/WSEAS International Conference on Heat Transfer, Thermal Engineering and Environment (HTE'08), Rhodes, Greece, August 20-22, 2008

On Electrocoagulation and Novel Hybrid Coagulation Method for Enhanced NOM Removal in Drinking Water Treatment.

Authors: Ondřej Gebouský[a,b], Jan Haidl[a], Michaela Prokopová[a], Martin Pivokonský[a]

[a] Institute of Hydrodynamics of The Czech Academy of Sciences, Pod Paťankou 30/5, Prague 6, Czech Republic

[b] University of Chemistry and Technology Prague, Technická 5, Prague 6, Czech Republic

Motivation

Natural organic matter (NOM) is a complex mixture of organic substances commonly found in surface and ground waters. Inherently, NOM is not toxic, but its presence in drinking water sources can significantly impact water quality, impair the efficiency of the drinking water treatment processes, and lower biostability in the supply network. On top of that, serious health risks are posed by the products of NOM reactions with disinfection agents – disinfection by-products, DBPs. Therefore, efficient NOM removal is paramount in drinking water treatment facilities for which various separation processes were developed and employed. However, with the ongoing global change, the composition of NOM in surface water alters, the levels increase, and their separability decreases. This calls for developing enhanced separation methods to improve the performance of existing facilities and alleviate the advanced (and expensive) separation techniques employed in any new facility. In the response, we present a novel hybrid coagulation method for NOM removal comprising conventional chemical coagulation (CC) and electrocoagulation (EC) combined into one enhanced separation step easily integrable into existing water treatment facilities.

Introduction

In the water treatment industry, both CC and EC are basic separation techniques based on destabilizing colloidal suspensions by adding metal coagulants and their consequent separation as precipitates. Both coagulation methods utilize a metal cation – usually Al^{3+} or Fe^{3+} - as the coagulation agent; however, they differ in the cation introduction. Whilst CC achieves this by adding a corresponding water-soluble salt, EC is based on a direct electrolytic dissolution of a metal electrode. The EC was proven to be a comparable or more efficient coagulation method for various pollutants in wastewater treatment, leveraging easier operation control, higher coagulant utilization, and lower sludge production than a conventional CC. For drinking water treatment, EC is less frequently used; it is utilized mainly for mobile water treatment units or removing heavy metals from ground waters [1]. Although few studies report the successful use of EC for NOM removal, the data are not compared to conventional CC, and coagulant dosages are mostly higher than acceptable for industrial use.

To develop a method for enhanced NOM removal and fill the knowledge gap in EC efficiency, we performed extensive experimental research comparing the efficiency of CC and EC for NOM removal from natural surface water. This work follows our previous research on optimizing chemical coagulation conditions for NOM removal [2]. The experiments were performed with Al- and Fe-based coagulants under optimal coagulation pH. Based on the obtained results, a hybrid coagulation method was developed, combining the economical treatment via CC with the efficiency of EC for effective and sustainable NOM removal.

Experimental setup and methodology

The experiments were conducted with natural surface water from the reservoir Vrchlice (Czech Republic), a source for the near-located DWTF. The raw water contains ~10 mg/L of dissolved organic carbon (DOC), of which almost 40% cannot be removed by conventional CC using either ferric or aluminum-based coagulants. The CC experiments were performed according to the standardized methodology using ferric sulfate and aluminum sulfate dosing from 4 mg/L to 40 mg/L of the metal at the pH optimized for each coagulant - 5.5 for $Fe_2(SO_4)_3$, 6.8 for $Al_2(SO_4)_3$. Before the coagulant addition, the water pH was adjusted with HCl/NaOH solution to the desired level; then, the coagulant was dosed, and the jar was stirred at a mean shear rate of 200 s^{-1} for 60 seconds. After the rapid mixing period, the mixing intensity was lowered to the shear rate of 40 s^{-1} for 30 minutes of the flocculation stage.

The EC experiments were conducted in a 2L stirred jar housing two cylindrical electrodes with 32 cm^2 active surface area. The aluminum and mild steel electrodes were used as anodes (sacrificial electrodes); the cathode was made of stainless steel. The electrodes were connected to a DC power supply unit operated in a constant current regime with a current intensity of 0.10 A. The picture of the experimental setup is shown in Figure 1. Similar to CC experiments, the EC experiments started with a pH adjustment, after which the electrolytic dissolution of the sacrificial electrode was performed in a jar stirred at a mean shear rate of 40 s^{-1}. The metal dosage from 1 mg/L to 40 mg/L was used; the duration of the dissolution step was calculated based on the calibration experiments. After the coagulant dosage, the electrodes were removed from the solution, and the jar content was mixed for an additional 30 minutes to allow for aggregate growth.

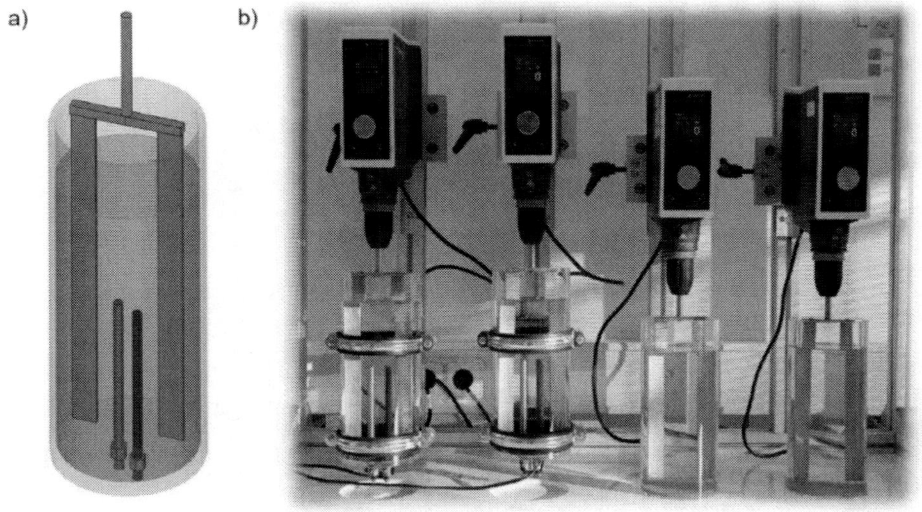

Figure 1. Experimental setup: a) Sketch of a standard stirred jar with EC electrodes. b) Photography of the experimental setup; left: jars for EC/DC, right: jars for CC.

The hybrid CC-EC experiments were performed, starting with standard CC conditions using an optimal dosage of coagulants determined in the previous investigation (8.5 mg/L of Fe or 12.5 mg/L of Al) [2]. After 10 minutes of the flocculation stage, the secondary coagulant (1 mg/L – 20 mg/L) was dosed using EC, and the jar content was left stirred for an additional 20 minutes of the flocculation stage.

After all the flocculation experiments, the solution was left for 15 minutes to settle; then, the sample of 250 mL was taken, centrifuged for 20 minutes at 4000 rpm, and analyzed for final pH, TOC concentration, and residual metal content.

Results and discussion

First, the potential of CC and EC for DOC removal was investigated separately using either Fe and Al sacrificial electrodes or $Fe_2(SO_4)_3$ and $Al_2(SO_4)_3$ as chemical coagulants. The results are shown in Figure 2 as the dependence of residual DOC concentration on the metal coagulant dosage. In all cases, the residual DOC concentration follows the Langmuir adsorption model proposed by Edwards [3],

$$\frac{c_{\mathrm{TOC}}^{0} - c_{\mathrm{TOC}}}{c_{\mathrm{M}}} = \frac{K(c_{\mathrm{TOC}} - c_{\mathrm{TOC}}^{\infty})}{1 + K(c_{\mathrm{TOC}} - c_{\mathrm{TOC}}^{\infty})}. \tag{1}$$

In the cases of aluminum coagulant, the fraction of non-purgeable DOC is almost identical, regardless of the coagulant addition method. On the other hand, in the case of EC performed with the mild steel sacrificial electrode, the fraction of non-purgeable DOC is significantly lower, and the coagulant utilization is higher than in the cases of Al-based coagulation and Fe-based CC – consult the model parameters listed in Table 1 - offering a promising method for DOC removal enhancement.

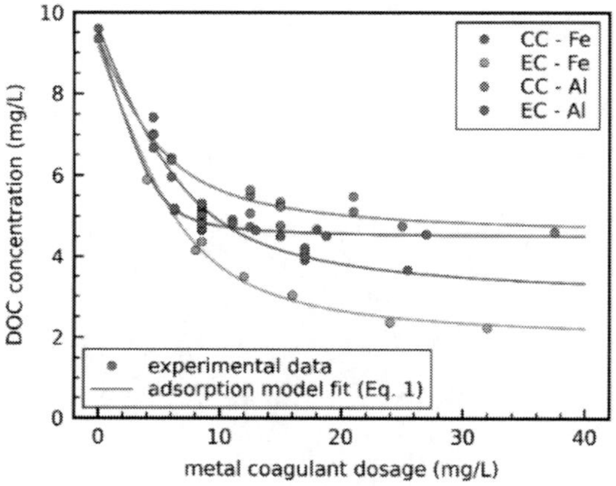

Figure 2. Residual DOC concentration as a function of the amount of coagulant added by CC and EC. Solid lines represent the fitted equilibrium adsorption model (Eq. 1).

Despite the substantially higher potential of EC with iron sacrificial electrodes for NOM removal in drinking water treatment, technical and economic difficulties limit its wide industrial utilization. While in industrial wastewater treatment, EC is proven to be an effective and sometimes cheaper alternative method to CC for the removal of various pollutants, the combination of low drinking water conductivity and its high production rates affects EC economy and technical feasibility. This usually prevents a full replacement of CC with EC as generating tens of mg/L of metal at the common production rates of hundreds to thousands of L/s becomes too expensive. However, these limitations can be partially overcome by combining the CC and EC into a hybrid process, employing EC as a secondary coagulant dosage method with relatively low coagulant dosage.

Table 1. Parameters of equilibrium adsorption model (Eq. 1) for four coagulation methods.

	CC-Al	EC - Al	CC - Fe	EC - Fe
c_{TOC}^{∞} (mg L^{-1})	4.5	4.5	2.9	1.9
K (L mg^{-1})	0.53	2.9	0.45	0.64
σ_c (mg L^{-1})	0.28	0.07	0.28	0.21

This work combined CC and EC into one treatment step, scheduling EC dosing 10 minutes after the CC. The optimal dosage of chemical coagulant determined in [2] was followed by 2.5 mg/L and 8.5 mg/L of metal dosed via EC. Figure 3 compares the DOC removal efficiency of the hybrid coagulation method to that of a conventional CC. It is shown that with only 2.5 mg/L of Fe dosed via EC, the DOC residual concentration is decreased significantly. In the case of the Fe-Fe coagulation system, the DOC reduction of an additional 1.1 mg/L is achieved; using only CC, such an efficiency would be achieved using more than 20 mg/L of Fe in total. In the Al-Fe and Al-Fe coagulation systems, a slightly lesser reduction of 0.9 mg/L in DOC is achieved; however, the resulting DOC concentrations are at or below the minimal values achievable with conventional Al-based chemical coagulation regardless of the coagulant dosage. With Fe-Fe and Al-Fe coagulation systems dosing 8.5 mg/L of Fe via EC, the resulting DOC concentration is always significantly lower than it can be achieved with CC using up to 40 mg/L of metal.

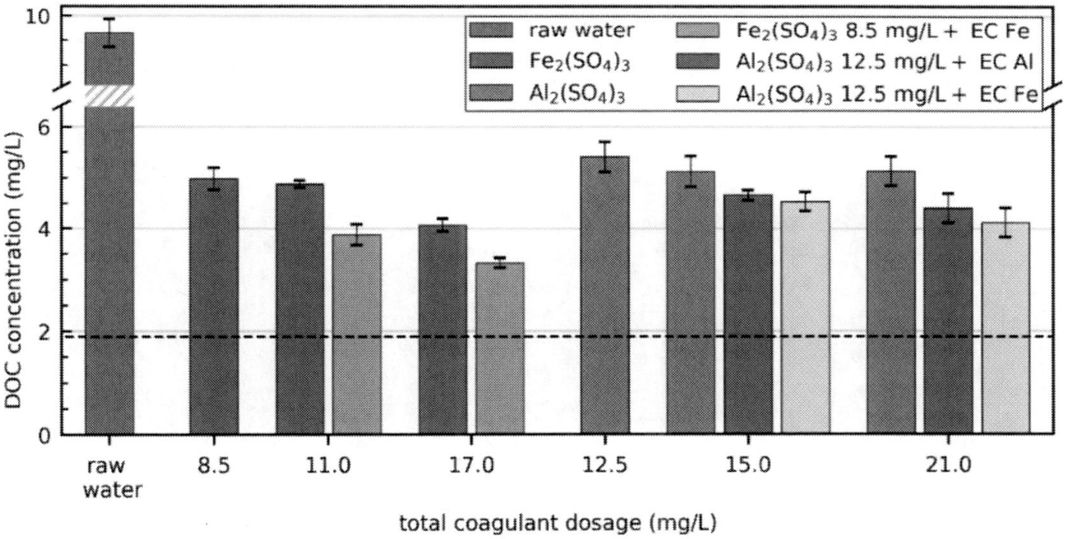

Figure 3. Comparison of the residual DOC concentration after conventional CC and hybrid CC-EC process.

Conclusions and implications

The EC is a more effective NOM removal coagulation method than the CC, which uses the same coagulant dosage. However, the EC's technical and economic aspects in drinking water treatment limit it from fully substituting CC. To overcome these difficulties and leverage the EC's efficiency, a hybrid coagulation method is developed and tested, replacing the CC only partially. The efficiency of the developed hybrid method is tested on natural surface water, for which its superiority is evidenced. The method can be relatively easily integrated into existing water treatment facilities, offering a simple method for NOM removal enhancement.

References

[1] Mousazedeh, M. et al. (2021). *Environmental Science and Pollution Research* 28(32), 43143.

[2] Pivokonsky, M. et al. (2024). *Journal of Water Process Engineering* 57, 104660.

[3] Edwards, M. (1997). *Journal AWWA* 89(5),78.

735bd To Investigate the Effect of Small Molecule Dopants (SMDS) on the Synthesis/ Electrical /Mechanical/Electro-Mechanical/Self-Healing Properties of PANI/Paampsa System

Arya Ajeev, Evan Wujcik

Stretchable electronic polymer (SEP) sensors have attracted significant interest due to their distinctive properties and versatile applications in the field of healthcare, artificial skin, human-machine intelligence etc. The development of lightweight, self-healable, stretchable, and high functioning components stands as a major requirement in wearable electronics, in consideration of the rigid metal and metal oxide-based sensors. In this work, the SEP system is composed of polyaniline (PANI), poly(2-acrylamido-2-methyl-1-propanesulfonic acid) (PAAMPSA) and small molecule dopants (SMDs). The polymer system is prepared through the oxidative polymerization of aniline while, A-PAM acts as a template to guide the PANI polymerization. SMD and A-PAM acts as dopants and cross-linking agents. Investigating the effect of SMDs molecular weight, structure, and the number and class of acidic groups (SO_3H, COOH, PO_3H) will provide insights into how these factors influence the intermolecular, thermal, morphological, self-healing, and electro-mechanical properties of the PANI/PAAMPSA system. The hydrogen bonds and electrostatic interactions between A-PAM, SMD, and PANI synergistically construct a homogeneous regenerative network, which contribute to the elasticity and soft compliant nature of material with extremely high stretchability. Among the various SMDs investigated, 4-dodecyl benzene sulfonic acid, has shown a conductivity of 0.07 S/m and a Young's Modulus of 39.76 kPa. Furthermore, SMDs such as pyrophosphoric acid, trifluoromethane sulfonic acid, and 5-sulfosalicylic acid dihydrate exhibited a higher stretchability, reaching up to 4000%.

735be Metal-Free Precision Bottlebrush Polymers with Ionic and Fluorinated Moieties for MRI Applications

Nduka Ogbonna, Titilayo Oluwole, Michael Dearman, Chamberlain Amofa, Yaxin An, Bhuvnesh Bharti, Jimmy Lawrence

Water-soluble fluorinated MRI contrast agents present safer alternatives to metal-based platforms for disease diagnosis. However, enhancing the fluorine signal compromises aqueous solubility. In this talk, we will discuss the design and synthesis of a new fluorine-rich water-soluble bottlebrush polymer featuring sequence-defined ionic side chains. The key to this synthetic approach is precise control over the side chains' sequence, composition, and structural dispersity. Our design concept underlines the importance of precision synthesis for enhancing NMR signal-to-noise ratio, signal sharpness, and MR image contrast. Such precision synthetic materials offer a safe and promising solution for future disease diagnosis and therapy.

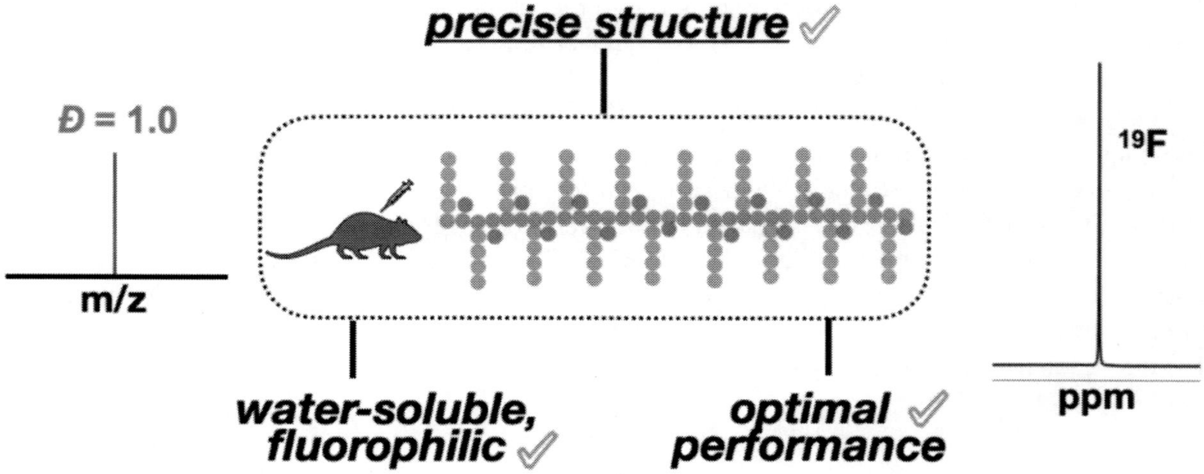

735bf Streamlining Bone Scaffold Production: Automation and Innovative Approaches

Yasser Ahmed, Ryan Striker, Hesham Abdelaziz, Ali Alshami

Bone scaffold fabrication is a major pillar in tissue engineering, yet scalability and reproducibility are challenges for conventional fabrication methods. This study aims to propose an innovative experiment design to leverage automation and innovation to enhance bone scaffold fabrication. 3D printing, material handling, and process optimization are components that achieve efficiency, consistency, and scalability improvements. Precise measurements were implemented for material handling challenges. A reduction in material waste was seen as a result. A mesh filter was developed to enhance scalability. The Filter enabled experimentation with multiple scaffolds while maintaining individual identity and measurements. This increased throughput, streamlined workflow, and improved efficiency. The highlight of our methodology is integrating automated casting with bioprinting technology. Through these improvements, higher precision and reduced processing time were attained. With the automated casting process, we were able to achieve over 5 times faster fabrication compared to manual methods, demonstrating scalability and efficiency gains. The focus is to produce high quality scaffolds, leveraging automation and innovation for advancing scaffold fabrication. This paper presents the experimental procedure in detail, the emphasis is on material handling improvements, as well as scalability solutions, and integrating automated casting. The end goal is optimizing the scaffold fabrication process, to advance tissue engineering.

735bg Adsorption of PFAS Using Ion Exchanged BEA Zeolite

Charles Ponge, Nathaniel Sheehan, David R. Corbin, Justin Hutchison, Edward Peltier, Mark B. Shiflett

Per- and Polyfluoroalkyl substances(PFAS) widely produced since the 1940s have a broad range of uses in consumer and industrial products. The persistence of PFAS in the environment, and the associated health risks, have called for monitoring and remediation of PFAS in drinking water in the US and around the world. Zeolites have been shown to be effective sorbents. While traditional sorbent materials such as activated carbon and ion exchange resins are being used to remove PFAS from municipal drinking water, the materials are not easily regenerated in situ. Zeolites have been identified as sorbent materials with the potential to be regenerated on location and demonstrate selectivity for sorbing PFAS. Zeolites have a wide range of pore openings and tunable properties, making them ideal for removing both short- and long-chain PFAS from the environment. Currently, there is minimal research investigating zeolites for sorption of PFAS. The ion exchange of zeolites has been shown to change the properties of zeolites to increase catalytic or adsorptive capability. In this presentation, experimental data for ion exchanged zeolite CP814E will be shown and how the adsorption properties change with varying metals, and different loadings. The adsorption of PFAS is generally increased after ion exchange from the hydrogen form, with some metals showing 50-100% increase in adsorption in HPLC water. Thermal regeneration of these zeolites is also shown to recover most of the adsorptive capacity.

735bh Iron Magnesium Coating Chicken Bones for Removal of Copper from Water

Mayar Tarek

Because heavy metals have detrimental effects on aquatic life and public health, their presence in freshwater sources is extremely concerning. Even in low quantities, copper (Cu), one of the most hazardous heavy metals, can result in major health problems. The current study seeks to create a novel adsorption medium using iron (Fe)-coated chicken bones Using magnesium hydroxides, copper can be eliminated from water. In which the chicken bones were treated with stacked double hydroxides (LDH) Physical adsorption is the process by which heavy metals from the waste liquid are adsorbed onto the chicken bone-based adsorbent. The cause of physical adsorption is an imbalance in the porous nature of the adsorbent surface, which draws the adsorbate (heavy metal) to the surface. The residue of inorganic materials left over after burning at temperatures higher than 575°C is known as ash. A substance is heated as part of the ash-making process, and the method involves drying, burning charcoal, and desolating volatile chemicals. Heat will be produced by burning the charcoal, volatile chemicals will evaporate, and moisture will be lost throughout the drying process. There will be an interaction between carbon and oxygen during the burning of charcoal to form co-gas, which then develops into $CO2$ gas so that in the end there will be only material known as ash. With a rough and porous surface shape, apatite carbonate compound minerals are a distinguishing feature of the adsorbent made from chickenbone ash. The adsorption performance can be improved by raising the ashing temperature, contact time, particle size, and weight of the adsorbent. The physical adsorption process is accelerated at near-neutral pH values. Adsorbents made of chicken bone ash can lower Cu metal levels while also improving waste quality.
In this paper, Carbon-based adsorption is described as a traditional treatment with a high unit cost, particularly in underdeveloped nations.

Contamination of metals, in water is a pressing concern for both the environment and public health. Several metals, including lead, mercury, arsenic, cadmium, and copper pose risks. These metals can find their way into water sources through processes well as industrial activities, mining operations, and agricultural runoff. Consuming drinking water with levels of metal contamination can result in health issues such as neurological problems, developmental disorders organ damage, and an increased risk of cancer. Furthermore, aquatic ecosystems are also adversely affected by metal contamination which impacts the well-being of fish and other aquatic organisms. Therefore it is crucial to address and mitigate metal contamination, to protect health and preserve the environment. Increased production of broiler chicken meat seems to be a result of Egyptians' rising consumption, which will lead to a rise in the amount of discarded chicken bones that cannot be recycled. Crystalline salt minerals in the form of calcium phosphate complexes, which are frequently idealized as hydroxyapatite, make up the majority of the inorganic phase of bone.Because the hydroxyapatite mineral has a porous surface and is a bioceramic material, it can function as an adsorbent. Metals in liquid waste can be adsorbed by this adsorbent, among its other uses. Liquid waste is produced by the manufacturing process, cleaning of production equipment, laboratory operations, and residual products that are deemed to be of low quality or that result from production process failure. among other sources within the petrochemicals industry sector. One alternative way is using waste treatment sorption. Adsorption using carbon is stated as a conventional treatment which has a high unit cost, especially in developing countries. This research was conducted to examine the possibility of recycling chicken bones for use in water treatment by coating the bones with iron and magnesium hydroxide layers.This new adsorption medium was applied to remove copper from a water sample.

Thus, this adsorption medium could be an eco-friendly and economically efficient alternative to the current methods for the removal of heavy metals from water.

735bj Fired Heater Modeling Using Htri

Avinashkumar Karre

A fired heater is the industry's largest energy consumer and NOx / CO2 emitters process. Furnace models are useful for troubleshooting plant problems, evaluating proposed changes to revamp old heaters for a new service, evaluating the addition of an economizer/air pre-heater to improve plant energy efficiency, evaluating the effect of proposed changes in plant operating conditions on furnace operation, estimating retirement life of tubes based on past and projected operations, and assist in assessing vendor furnace designs. This paper shows a stepwise modeling of a fired heater using HTRI software. The rigorous model utilizes four furnace components:

- Combustion module:
  - Any fuel type(s) (liquid, gas)
  - Any oxidant stream (e.g., air, turbine exhaust)
  - Specify %XS air or O2% in stack
- Radiant coil module (API 530 method):
  - Tube thickness, radial temp profiles, tube life, heat flux (circumferential)
- Convection section module:
  - Duty, flue gas, and metal temperatures calculated
  - Draft profiles
- Radiant section module:
  - 3-D Hottel Zone method for radiation heat transfer calculations
  - Integrated rigorous process side calculations (mass, heat, momentum transport)
  - Multiple burners modeled as superimposed individual jet flows

735bl Optimizing Vaccine Distribution Using Attainable Region Theory

Neil Thomas Stacey, Diane Hildebrandt, Baraka Celestin Sempuga, Klaus Moller

The emergence of COVID-19 presented a unique global challenge, with a race to develop vaccines and distribute them rapidly and equitably. A range of vaccines were developed at an unprecedented pace, but bottlenecks of manufacturing and distribution obstructed their optimal use in the global context. The experiences of COVID-19 clearly show that to be prepared for future pandemics, methods are needed to select optimal vaccine strategies within the constraints of manufacturing capacity and cost so that supply chains can be developed to support this.

Attainable Region Theory, a field of mathematics developed initially for the design and optimization of chemical reactor networks, provides a systematic framework for this task, providing the ability to visualize different vaccination strategies, both homologous and heterologous, and combinations of those strategies. In the same way that Attainable Region Theory allows a reactor designer to map the hyper-volume of possible process outputs into a two-dimensional space of Enthalpy and Gibbs Free Energy of reaction, it also offers a means of mapping the hyper-volume of all feasible vaccination strategies into a two-dimensional space of total preventive efficacy and cost, within the constraints of available manufacturing capacity.

We demonstrate that this approach allows for rapid comparisons of vast sets of possible strategies at both local and global scales and allows for the projection of those strategies over multiple time scales. This approach allows for rapid optimization for utilizing all available resources, showcasing the potential of Attainable Region Theory to inform data-driven, cost-effective, and adaptable vaccine distribution strategies. By optimizing the deployment of vaccines, including heterologous approaches, we can develop supply chains that support global vaccination, ultimately enhancing public health outcomes.

735bm Machine Learning Predictive Modelling of Velocity and Turbulent Kinetic Energy of a Bladeless Wind Turbine Implementation in an Urban Environment

Abdulfatai Faro, David Cheneler, Marianna Cavada, Aonullahi Adebayo

The installation of wind turbines to all possible extents has contributed to meeting the ever-increasing energy demand. Technologically developed areas have a high potential for wind energy, including the rooftops of high-rise buildings, railway tracks, the region between or around multi-story buildings, and city roads. Harnessing wind energy from these areas is quite challenging since it has a dramatic nature, being chaotic and turbulent on urban surfaces. A modelling paradigm is developed to assist predictive models of two turbulence models; SST and RNG, by using generated data from computational fluid dynamics (CFD) simulations. The key components of the current study involve four machine learning techniques validated to estimate the velocity and turbulence kinetic energy of a wind pattern in an urban environment: support vector regression, artificial neural networks (ANN), random forests, and extreme gradient boosting (XGBoost). To create training and test datasets for the four machine learning algorithms, a series of high-fidelity numerical simulations for a bladeless wind turbine implementation in an urban environment were performed; The capability of this bladeless wind turbine can harness winds from any direction. The ML models' anticipated velocity and turbulence kinetic energy are also compared to key existing analytical velocity and turbulence kinetic energy models.

The machine learning methods estimate velocity and turbulence kinetic energy in a manner comparable to CFD simulations. The findings show that machine learning-based systems can estimate velocity more accurately than traditional methods. The broader vision is that a comparison study of these turbulence models was established against machine learning techniques to reveal the form of the models' discrepancy and improve the predictions of data-driven turbulence models not just for wind patterns but for more general flows.

735bn Template-Based Sentence Generation for Variable Definition Extraction from Papers on Chemical Processes

Kotaro Nagayama, Shota Kato, Manabu Kano

Introduction

Physical models are crucial in designing processes and optimizing operational conditions in the process industry. Process simulators based on physical models exist for some processes, but they do not apply to all processes. When dealing with a process for which no suitable simulator is available, it is necessary to construct a new physical model. However, constructing physical models that can accurately replicate the behavior of the target process requires a thorough review of existing literature to gather and assimilate a substantial amount of information. To facilitate this labor-intensive task, we aim to realize Automated Physical Model Builder (AutoPMoB), which automatically builds desired physical models from relevant documents [1]. A key component of this system is a method for accurately extracting variable definitions from documents.

In the task of variable definition extraction, methods employing pre-trained models, such as BERT (Bidirectional Encoder Representations from Transformers) [2], have shown promising results. Yamamoto et al. [3] achieved the highest performance in extracting variable definitions from chemical process-related papers using a BERT model. Their method takes a sentence with a target variable and replaces the target variable with a special token. The sentence is then inputted to the BERT model and the position of the corresponding variable definition is obtained. They constructed a definition extraction model through a two-step fine-tuning process, first using the Symlink dataset [4] comprising five fields with low relevance to chemical processes, followed by a dataset related to chemical processes, achieving an accuracy of 85.5%. However, since variable definition extraction is an upstream process in AutoPMoB and influences the overall performance, its performance should be nearly perfect. From this perspective, the method developed by Yamamoto et al., despite its achievements, falls short of the necessary standards for AutoPMoB.

To enhance the variable definition extraction performance, we focus on the size of the training data used by Yamamoto et al. The size is smaller than that used in related studies; thus, the performance can be improved by increasing the training data. However, since the characteristics of variable definitions, such as their lengths and the words they contain, differ across fields, the difficulty of variable definition extraction also varies by field [5]. Therefore, increasing training data in other fields may not improve the performance. Besides, generating substantial new data for each field requires a high cost. To overcome these obstacles, we propose a method of expanding the training data in this study.

Proposed Method

The diagram in Figure 1 outlines our proposed approach. Initially, we prepare a set of template sentences using ChatGPT, such as "[VAR] is defined as [DEF]." The prompts for ChatGPT include detailed instructions on the desired output format, the number of variables in the sentence to be generated, and examples. The proposed method generates new definition sentences by substituting variables and definitions from existing training data into these template sentences. The generation of definition sentences continues until all variables in the training data are used.

Results and Discussion

We evaluated our model, trained with the training data augmented by the proposed method, against the model developed by Yamamoto et al., using a dataset created from 47 papers on five chemical processes. Our model achieved an accuracy of 89.6%, which surpassed Yamamoto et al.'s model by 2.8 percentage points. Notably, our model maintained high accuracy even without using data from the test for training, a scenario where the existing method's performance typically declines. Although overall definition extraction performance was improved, there remain cases where both existing and proposed methods fail, such as sentences with multiple variables, definitions, and equations. The template sentences created in this study were more straightforward than the actual definition sentences; for example, they contained multiple [VAR]s and [DEF]s, but all of them were mapped to each other, and no equation was included in the templates. Further performance improvements can be expected by developing a method of creating template sentences that more closely resemble the definition sentences appearing in the dataset.

Acknowledgment

This work was supported by JSPS KAKENHI Grant Number JP23K13595.

References

[1] Kato, S. and Kano, M.: Towards an Automated Physical Model Builder: CSTR Case Study, Computer Aided Chemical Engineering, vol. 49, pp. 1669–1674 (2022)

[2] Devin, J., Chang, M.W., Lee, K., and Toutanova, K.: BERT: Pretraining of deep bidirectional transformers for language understanding. In: Proceedings of the 2019 Conference of the North American Chapter of the Association for Computational Linguistics: Human Language Technologies, Volume 1 (Long and Short Papers). pp. 4171–4186 (2019)

[3] Yamamoto, M., Kato, S., and Kano, M.: Variable definition extraction by BERT with two-step fine-tuning. In: Proceedings of the 29th Annual Conference of the Association for Natural Language Processing. pp. 2957–2961 (2023)

[4] Lai, V.D., Veyseh, A.P.B., Dernoncourt, F., and Nguyen, T.H.: Semeval 2022 task 12: Symlink - linking mathematical symbols to their descriptions. In: Proceedings of the 16th International Workshop on Semantic Evaluation (SemEval-2022). pp. 1671–1678 (2022)

[5] Popovic, N., Laurito, W., and Färber, M.: AIFB-WebScience at SemEval-2022 task 12: Relation extraction first - using relation extraction to identify entities. In: Proceedings of the 16th International Workshop on Semantic Evaluation (SemEval-2022). pp. 1687–1694 (2022)

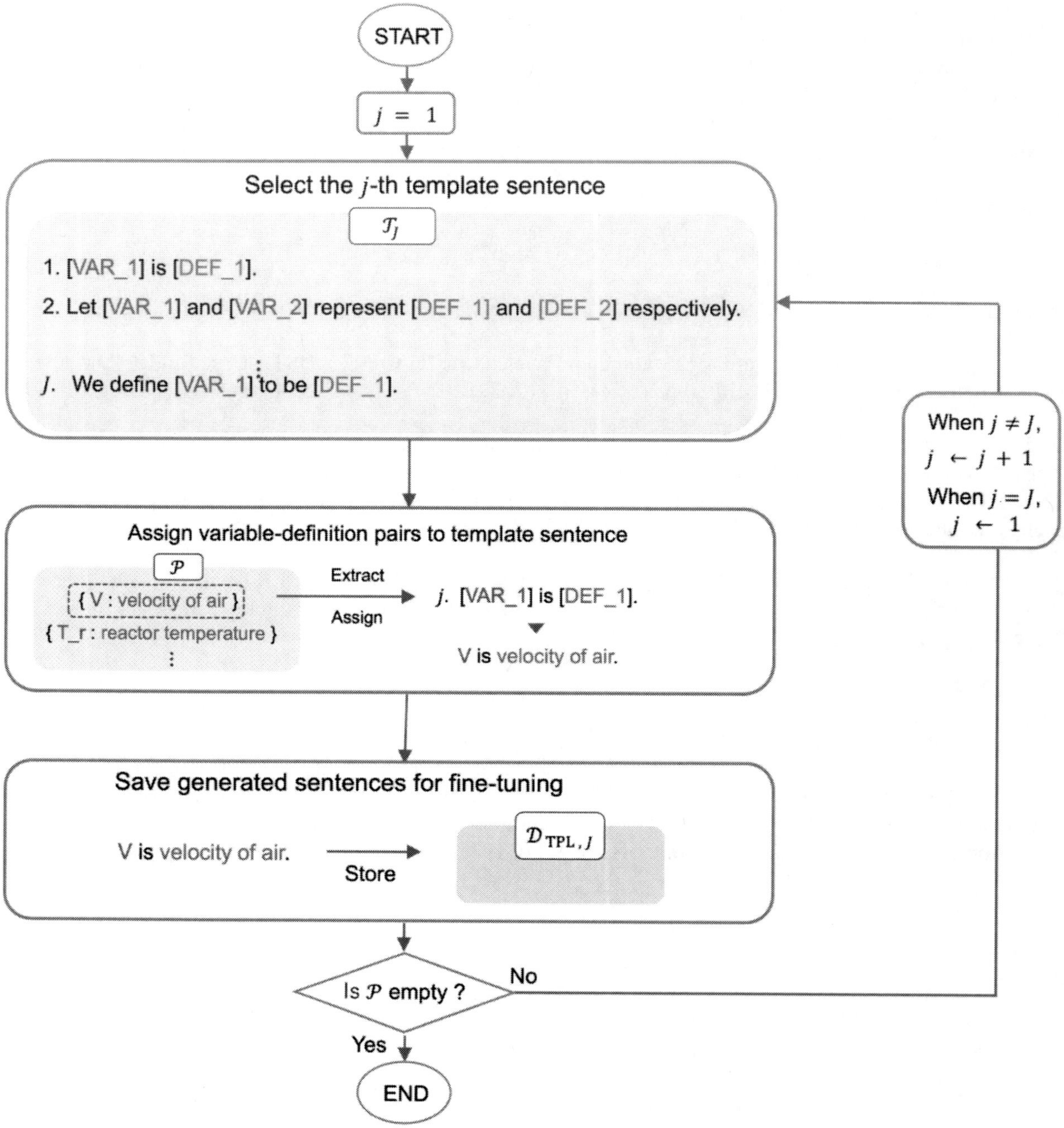

Figure 1: A schematic diagram of the proposed method.

735bo Probing RNAs and RNA Binding Proteins in the Pathogenesis of *Multiple Sclerosis*

Vailankanni L. Rodrigues, Gregory L. Dignon

Multiple Sclerosis is an autoimmune disease of the Central Nervous System that affects the brain and the spinal cord. The immune system attacks the insulating covering of the nerve fibers called myelin, causing inflammation and damage to the underlying nerves. Here, we look at these disease-causing RNA-binding proteins that are implicated in the pathogenesis of MS using computational methods. We particularly focus on the dysregulation of these proteins which leads to aggregation, mislocalization, and aberrant formation of stress granules.

735bp De Novo Design for Small Molecule Binding and Sensing

Linna An, Meerit Said, Long Tran, Sagardip Majumder, David Baker

A general method for designing proteins to bind and sense any small molecule of interest would be widely useful for high throughput assay development for enzyme engineering. Due to the small number of atoms to interact with, binding to small molecules with high affinity requires highly shape complementary pockets, and transducing binding events into signals is challenging. Here we describe an integrated deep learning and energy based approach for designing high shape complementarity binders to small molecules that are poised for downstream sensing applications. We employ deep learning generated psuedocycles with repeating structural units surrounding central pockets; depending on the geometry of the structural unit and repeat number, these pockets span wide ranges of sizes and shapes. For a small molecule target of interest, we extensively sample high shape complementarity pseudocycles to generate large numbers of customized potential binding pockets; the ligand binding poses and the interacting interfaces are then optimized for high affinity binding. We computationally design binders to four diverse molecules, including for the first time polar flexible molecules such as methotrexate and thyroxine, which are expressed at high levels and have nanomolar affinities straight out of the computer. Co-crystal structures are nearly identical to the design models. Taking advantage of the modular repeating structure of pseudocycles and central location of the binding pockets, we constructed low noise nanopore sensors and chemically induced dimerization (CID) systems by splitting the binders into domains which assemble into the original pseudocycle pocket upon target molecule addition. This work is currently under peer review. Till now, we further improved our CID design computational pipelines using more effective diffusion algorithms, and are currently implementing our designs as sensors in metabolic engineering for critical compounds, such as taxol.

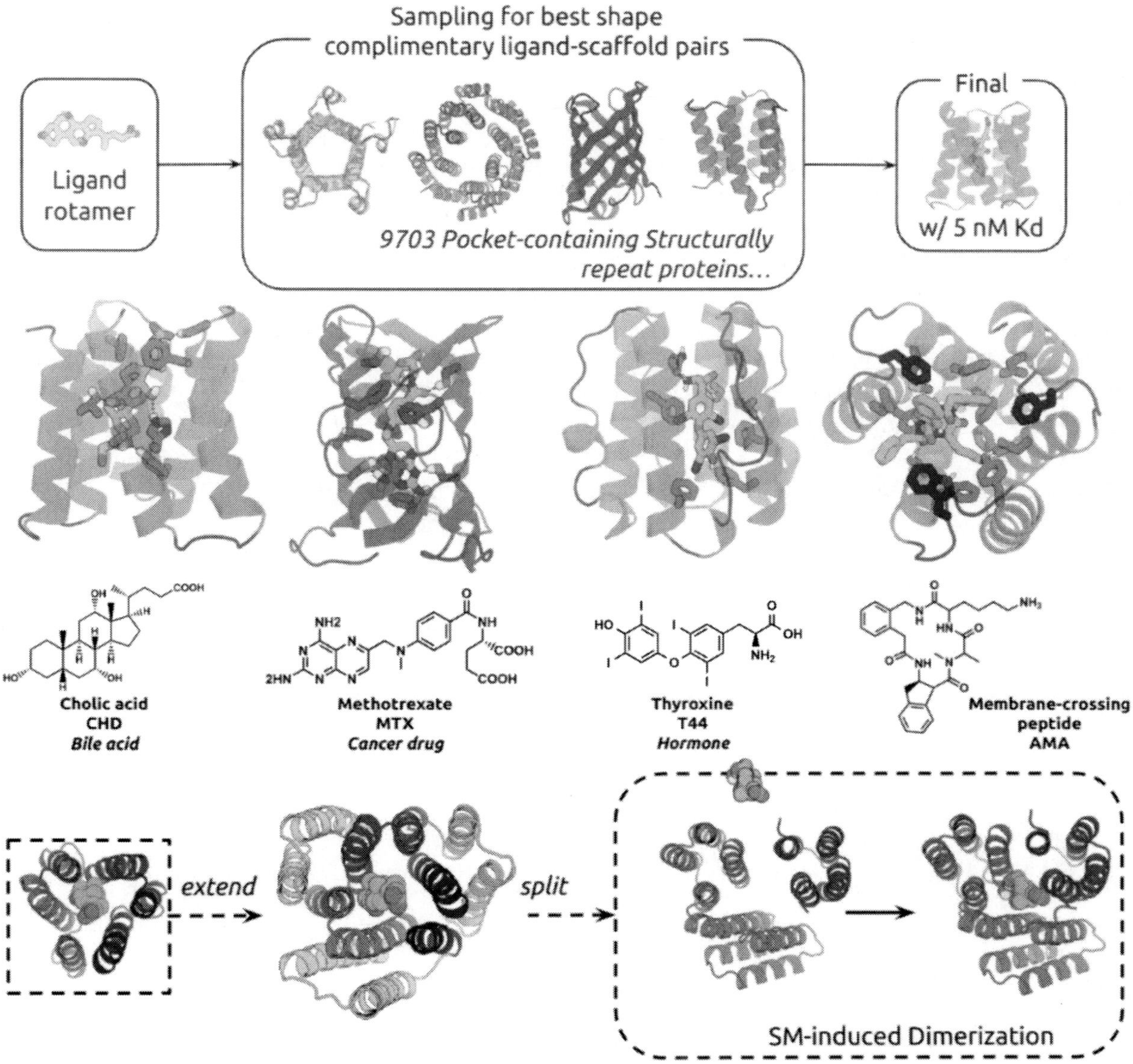
Sampling for best shape
complimentary ligand-scaffold pairs
Ligand
rotamer
9703 Pocket-containing Structurally
repeat proteins...
Final
w/ 5 nM Kd
Cholic acid
CHD
Bile acid
Methotrexate
MTX
Cancer drug
Thyroxine
T44
Hormone
Membrane-crossing
peptide
AMA
extend
split
SM-induced Dimerization

735bq Zinc Oxide Nanoparticles As Nanofertilizers: Effect on Morpho-Physiological Traits and Antioxidant Properties of Mustard (*Brassica juncea*)

De'Zarae Guthrie

The utilization of nanomaterials derived from essential nutrients such as zinc holds promise for the development of nanofertilizers and nanocomposites, which can find extensive applications in agriculture and human nutrition enhancement. This research aimed to produce zinc oxide nanoparticles (ZnO NPs) utilizing pecan (Carya illinoinensis) leaf extract and examine their impact on the growth, physiology, nutrient levels, and antioxidant properties of mustard (Brassica juncea) plants. The ZnO NPs underwent characterization through UV-Vis spectrophotometry, Dynamic Light Scattering (DLS), and Scanning Electron Microscopy (SEM). Mustard plants were exposed to varying concentrations of ZnO NPs (ranging from 0 to 200 mg L-1) during the vegetative growth phase. The DLS analysis revealed monodispersed ZnO NPs with a size of 84.5 nm and a highly negative zeta potential (-22.4 mV). Overall, the application of ZnO NPs led to significant enhancements in growth, chlorophyll content (increasing by 53%), relative water content (rising by 46%), membrane stability (improved by 54%), and net photosynthesis in a dose-dependent manner. Furthermore, supplementation with ZnO NPs elevated the levels of potassium, iron, zinc, and flavonoids, while also mitigating the impact of reactive oxygen species by boosting antioxidant capacity in mustard leaves by up to 97%. In summary, ZnO NPs show promise as a stimulant for plant growth and a novel soil amendment to increase crop yields. Additionally, biofortifying B. juncea plants with ZnO NPs could enhance the crop's nutritional value and potentially amplify its medicinal properties.

735bs Constructing Customized Pano-Structured Material Systems for Enhanced Sustainability and Health Monitoring

Yuanwei Li, Chad A. Mirkin, Jennifer A. Dionne

Synthetic oligonucleotides are crucial biomolecules that have profoundly impacted diverse fields such as material science, healthcare, and diagnostics. Their ability to form specific structures on demand and interact with a wide range of targets has opened new avenues for creating programmable and functional materials, enabling advancements in nanotechnology and healthcare. In this talk, I will discuss two specific examples utilizing synthetic oligonucleotides: (i) using DNA to engineer functional nanomaterials from anisotropic colloidal nanoparticles, and (ii) leveraging DNA aptamers and high-quality metasurfaces for multiplexed metabolite detection.

In the first example, I will explore the use of DNA and precision synthesis techniques to engineer programmable nanomaterials from anisotropic colloidal nanoparticles, such as polyhedral nanoparticles, nanoframes, and nanocages. Employing DNA as a molecular blueprint, I developed a universal synthetic strategy that generated a diverse library of designer nanostructures. This approach allowed us to precisely control the assembly of nanoparticles into colloidal crystals with tailored chemical, optical, and mechanical properties. Importantly, my work addressed significant synthetic gaps in porous crystal design over the 10-1000 nm length scale, enabling new opportunities for the loading and transport of large guests at this size regime. These advancements have broad implications, such as improving energy conversion systems, enhancing biodetection, and creating metamaterials with unique mechanical and optical properties, including negative refractive indices. Ultimately, these developments could contribute significantly to advancements in sustainability and healthcare.

In the second example, I will describe how different DNA sequences can be utilized to develop spherical nucleic acid (SNA)-based chemical probes for biomolecular detection and monitoring. These SNAs, which consist of densely packed nucleic acids arranged spherically around a nanoparticle core, are integrated with high-quality (high-Q) dielectric metasurfaces for the multiplexed detection of metabolite biomarkers related to chronic stress, such as adenosine, dopamine, oestradiol, and cortisol. Specifically, split DNA aptamers are attached to SNA nanoparticle cores and resonant silicon nanophotonic antennas to establish an ultrasensitive sandwich detection mechanism. The DNA aptamers on metasurfaces and SNAs interact with target analytes, forming a secondary DNA structure that localizes SNAs and analytes on dielectric metasurfaces. This creates significant optical scattering changes linked to biomarker concentrations and causes a visible color change indicating biomarker presence. This innovative approach allows for the detection of picomolar concentrations of multiple biomarkers simultaneously, providing crucial health metrics from biological samples like sweat. Moreover, our modular system can be adapted for detecting various other molecules, including proteins, microRNAs, and ions, showcasing its versatility and potential for broad applications in precision health monitoring.

Compressor Surge Control Line Mapping

Saleh Al Musallam

An integral milestone of centrifugal compressor commissioning is to perform surge testing in order to accurately determine the real surge points. Centrifugal compressors are widely used within gas processing sector within the energy industry. Unavailability of this equipment due to premature failure will result in revenue loss, and escalated maintenance cost. An important contributor to machinery interruption and unstable operation caused by excessive surging. Additionally, gas compressors are consuming significant amount of energy and improper surge mapping during plant operation would lead to energy losses. Surge testing establishes the baseline for compressor curve for equipment protection, safe operation and enhanced process efficiency.

This paper highlights surge test procedures implemented during the commissioning of centrifugal gas compressor of accurate mapping of surge control line during plant initial startup.

735bt Fluorescent Nanothermometry Enhanced Laser Tissue Soldering for Minimally Invasive and Robotic Surgery of Fragile Tissues

Oscar Cipolato, Tobias Leuthold, Marius Zäch, Georg Männel, Sam Aegerter, Calinda Sciascia, Alexander Jessernig, Sima Sarcevic, Jachym Rosendorf, Vaclav Liska, Dennis Kundrat, Romain Quidant, Inge Herrmann

Laser tissue soldering represents a promising alternative to sutures and staples: it is a surgical technique that combines laser light with a solder paste, a temperature-activatable adhesive material, usually protein-based, to create strong watertight bonds. Furthermore, this technique is, in principle, readily adaptable for minimally invasive procedures. This offers clear advantages over suturing and stapling, as these are difficult to perform in a minimally invasive setting and, when performed on fragile tissues like blood vessels, nerves, and intestine, are susceptible to inflammation, infection, and dangerous fluid leaks. However, the technique's success hinges crucially on the precise temperature management during the procedure.

In our work, we introduced a new approach based on controlled laser tissue soldering guided by fluorescent nanothermometers able to create seamless water-tight sealing. Nanothermometers have been successfully integrated into proteinaceous solders and offer non-invasive temperature measurements. As a critical step towards more accurate and less invasive soldering, we demonstrated automatic temperature control and smart laser power modulation for high-performance laser tissue soldering, enabled by an intelligent solder paste that contains NIR-fluorescent nanothermometers and nanoabsorbers, such as gold nanorods or titanium nitrate nanoparticles. Moreover, we present integrating computer-vision based solder recognition along with multiple approaches to guide the surgery by measuring the nanothermometry-based temperature distribution during soldering for optimal outcome. Finally, we demonstrated the crucial advantages of minimally invasive soldering in clinically-relevant scenarios.

735bx Protein-Based Isocratic Separation of Rees Using Newly Identified Calcium-Binding Proteins.

Farid Khoury, Piyush Kumar, Sameera Abeyrathna, Jonathan Picking, Brad Heater, Kate Kucharzyk, Scott Banta

Rare earth elements comprising the 15 lanthanides, yttrium, and scandium are critical for a green economy. Conventional chemical separation of these vital elements is complex, and the process is often compared to fine chemical processing, not commodity metals. Biological systems have evolved to uptake and utilize calcium, which shares similar physical and chemical properties with lanthanides, such as ionic radius and coordination geometry. We identified possible selective lanthanide-binding proteins via a bioinformatic search of calcium-binding proteins and modeling. The top 10 most promising candidates were expressed as a fusion of fluorescence resonance energy transfer (FRET) pair cyan fluorescent protein (CFP) and yellow fluorescent protein (YFP). The selectivity of the proteins to the ions within the lanthanide series were evaluated using FRET. The results improve our understanding of the effect of the size and chemistry of the binding pocket and the secondary structure of the protein as a whole on the ion selectivity. The repeat in toxin (RTX) and HEW5 domains had the highest selectivity, with the opposite preference for heavy and light REEs. The selected domains have high capacity with 8-10 binding lanthanide sites per protein (15 kDa) and are pH- and temperature-tolerant. The selected proteins were further characterized and used to separate rare earth elements (REEs) from one another and from non-REEs present in both primary and secondary sources with high yield and purity under isocratic chromatography conditions without the use of chelators.

This research was developed with funding from the Defense Advanced Research Projects Agency (DARPA). The views, opinions and/or findings expressed are those of the author and should not be interpreted as representing the official views or policies of the Department of Defense or the U.S. Government. Approved for Public Release, Distribution Unlimited.

735bz Removal of Organic Matter from Ceramic Wastewater Using Coagulation Process By Iron (III) Chloride

Ahmed Mahmoud, Mohamed Mostafa, Robert Peters, Shimaa R. Hamed, E. Khamis

Eliminating hazardous elements from industrial effluents is a challenging task for Egypt's industrial sector, especially when reusing the wastewater for the same industrial uses. One of these industries is the ceramics industry. This study successfully treated organic contaminants from ceramic wastewater using ferric chloride coagulation and flocculation techniques. The raw wastewater contained high levels of Chemical Oxygen Demand (COD), Biochemical Oxygen Demand (BOD), Total Suspended Solids (TSS), Total phosphorus (TP), Total nitrogen (TN), oil, grease, and hydrogen sulfide (H_2S). These levels were 8560, 4920, 5100, 2.8, 65.4, 120, and 6.2 mg/L, respectively, and the removal efficiency was 85.6, 82.5, 82.9, 53.6, 58.9, 92.6, and 62.9%. At an initial pH of 9, we found that the most effective treatment approach was to use a dose of ferric chloride 1 g/L for 60 minutes, followed by 300 rpm of agitation for 5 minutes, 30 rpm of floc production for 20 minutes, and 35 minutes of settling time.

AUTHOR INDEX

AIChE
120 Wall Street, FL 23
New York, NY 10005-4020

ISBN 979-8-3313-1689-1

CURRAN ASSOCIATES INC.
proceedings.com